A Guided Tour of the Living Cell

A GUIDED TOUR OF THE LIVING CELL

VOLUME TWO

Christian de Duve

Illustrated by Neil O. Hardy

This book is published in collaboration with
The Rockefeller University Press

**SCIENTIFIC
AMERICAN
LIBRARY**

An imprint of Scientific American Books, Inc.
New York

Library of Congress Cataloging in Publication Data

De Duve, Christian.
   A guided tour of the living cell.

   (Scientific American library)
   "This book is published in collaboration with
the Rockefeller University Press."
   Includes index.
   1. Cells.   I. Hardy, Neil O.   II. Title.
QH581.2.D43   1984   574.87   84-5534
ISBN 0-7167-5002-3 (v. 1)
ISBN 0-7167-5006-6 (v. 2)

Printed in the United States of America

Book design by Malcolm Grear Designers

Scientific American Library is published
by Scientific American Books, Inc., a subsidiary
of Scientific American, Inc.

Distributed by W. H. Freeman and Company,
41 Madison Avenue, New York, New York 10010.

1 2 3 4 5 6 7 8 9 KP 2 1 0 8 9 9 8 7 6 5 4

Contents

Volume One

A Guided Tour of the Living Cell

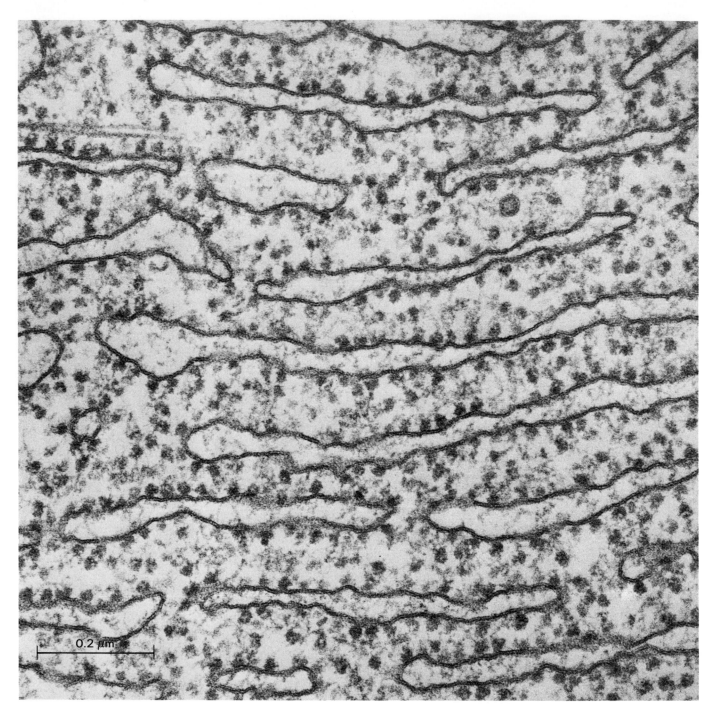

0.2 μm

In this section through a rough ER region of a guinea-pig-pancreas acinar cell, flattened cisternae are seen packed so close together as to leave only very narrow passages between them. Note the ribosomes attached by their large subunit (see Chapter 15) to the cytoplasmic face of the ER membranes. The two ribosomal subunits are clearly discernible in a number of the ribosome profiles. The fuzzy material inside the cisternae consists of secretory proteins.

13 | Membranes in Action: A View from the Cytosol

One of the most conspicuous features of many eukaryotic cells is the high degree of development of their cyto-membrane system. In some cells, the total surface area of this system may exceed 0.1 mm², which may not seem like much until you adjust it to our millionfold magnification. Then you find that it means dividing up the space of a large auditorium with more than one million square feet of partitions. Moving through such an auditorium may be a problem, as indeed it is in the ER-rich areas of the cell, where flattened cisternae may be packed so close together as to leave only very narrow passages between them. Fortunately, the partitions are flexible and can be forced apart fairly easily, as shown by the occasional mitochondrion or other cytoplasmic granule found nestling between ER cisternae.

Membranes as Organelles

During the first part of our tour, we were given plenty of proof that these membranes are not mere partitions. But relatively little of what they do, and especially of how they do it, could be discerned on their *trans* face. And so our descriptions had to remain largely phenomenological. Now that we can inspect the *cis* face of the membranes, we get a much clearer view of the number and variety of active systems that are compressed in these tenuous films. Even more important, we can try to find out how they operate, as their works are largely exposed on the side we are facing, which is where they are supplied with substrates, receive energy, interact with

Schematic representation of a permease. This drawing shows how the binding of a substrate causes a conformational change such that a channel is opened allowing transloca-tion of the substrate. Note that the system is perfectly reversible; transport takes place in the direction of decreasing (electro)chemical potential.

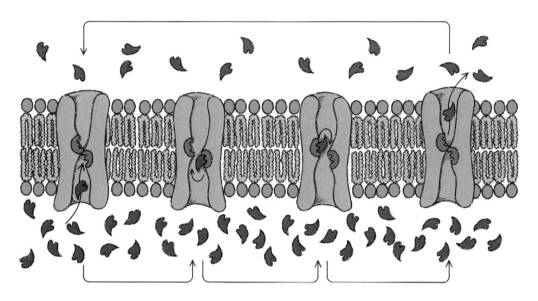

cofactors of various kinds, connect with cytoskeletal elements, and, in general, relate with other parts of the cell. Provided we can decode what is before our eyes, which is not yet possible for many of these complex machineries, we can now aim at understanding how all those displays are engineered. And understanding, not description, is the true purpose of scientific exploration.

Let us take another tour, therefore, through the meandering spaces of the cytosol and scan the surface of the membranes for evidence of the functions that they fulfill. These functions may be grouped under two main classes. One concerns all forms of exchanges—of matter, energy, information—between the two regions separated by the membrane. The other includes a variety of metabolic processes that have in common that they depend on, or take advantage of, some of the structural features provided by membranes, such as a hydrophobic milieu, a substratum for the creation of electric potential differ-ences, or a framework on which multiprotein complexes can be assembled and disassembled with great rapidity and accuracy.

Behind the Plasma Membrane

For a start, let us find out what lies behind the formidable array of gates, checkpoints, transport systems, sensors, and antennae of various kinds with which living cells greet their visitors (Chapter 3). What these systems have in common is a dependence on transmembrane proteins equipped with binding sites that allow the recognition of some specific chemical entity. What differentiates them, in addition to the kind of ligands they bind, is the way in which they respond to ligand binding. In terms of a frequently used analogy, the binding sites are locks, and the ligands are the keys that fit the locks.

Coupled facilitated transport of two solutes.

A. Symport: Both solutes are transported in the same direction.

B. Antiport: The two solutes are transported in opposite directions.

In both examples shown, downhill transport of the red molecules drives

uphill transport of the blue molecules (overall $\triangle G < 0$).

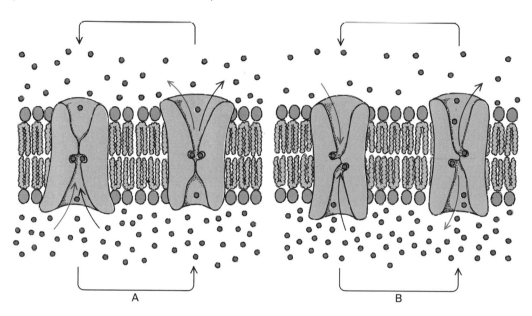

A

B

What follows the introduction of the key into the lock—the binding of the ligand to its binding site—varies from one system to another. Almost invariably, the first consequence is a change in the shape of the receptor molecule—the lock is turned—which either affects its function or activates some transducer or signaling device to which it is connected.

Molecular and Ionic Transport

One type of response is transport of the ligand across the lipid bilayer. This is what permeases do. Upon binding of their ligand on one side of the bilayer, they undergo a change in conformation such that the occupied site now faces the other side, where it can unload its molecular passenger. As was explained in Chapter 3 (see also Appendix 2), this mechanism requires no source of energy other than a difference in the chemical (or electrochemical) potentials of the transported substance across

the membrane. Accordingly, the transport is always in the direction of decreasing potential. Permeases, nevertheless, are very useful because they help vitally important hydrophilic substances, such as glucose or amino acids, cross what would otherwise be an essentially impassable barrier, and do so in a very specific fashion.

A number of permeases can translocate a substance only in combination with (symport) or in exchange for (antiport) some other substance. Such systems are of interest because they can move one of their substrates up against a potential gradient, provided the other substrate simultaneously moves down a steeper gradient. They link one flow to the other, the way a rope and pulley may couple the lifting of one weight to the fall of a heavier one.

Pumps are constructed like permeases, but with the fundamental difference that the translocating shift of occupied binding sites is driven forcibly with the help of energy. In the case of the all-important sodium-potassium

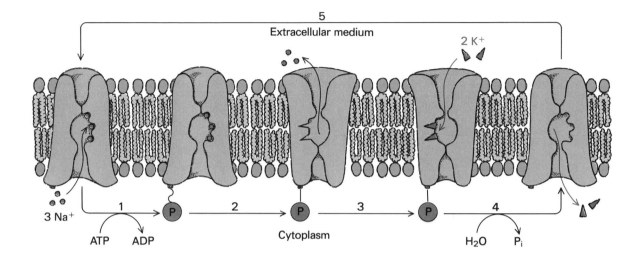

pump, the energy is supplied by ATP. When the sodium-binding sites are occupied on the *cis* side of the plasma membrane, a γ_d attack transfers the terminal phosphoryl group of ATP onto an aspartyl residue of the enzyme. This reaction conserves most of the γ-bond energy in the acyl-phosphate bond that is formed (Chapter 8). It causes the conformation of the protein to change in such a way that the sodium-binding sites now face the outside and, at the same time, lose their affinity for their substrate so that the bound ions fall off. Three sodium ions are expelled in this way against a combined electrochemical potential of the order of 3.3 kcal per equivalent (Appendix 2), at a total cost, therefore, of about 10 kcal per gram-molecule of ATP used. This energy is provided by a concomitant fall of the phosphoryl group in the enzyme from a high to a low potential level. Return to the initial state is mediated by the hydrolytic removal of this phosphoryl group. This process can take place only if two externally oriented binding sites are occupied by potassium ions and obligatorily results in the unloading of these sites inside the cell. As shown in Appendix 2, this step requires little energy, because the electric membrane potential compensates for the positive difference in chemical potential of intracellular over extracellular potassium ions. According to this description, the sodium-potassium pump may be defined in technical terms as an ATP-driven, vectorial, antiport system, which exchanges three intracellular sodium ions for two extracellular potassium ions per molecule of ATP hydrolyzed. Because the pump tends to deplete the cell of positively charged ions, it is electrogenic. Its function, however, resembles that of a bilge pump. It consists essentially in the correction of leaks and must be interpreted in the context of the various passive ion movements across the membrane. The sodium-potassium pump can even be made to move backward by high enough ionic gradients. It then powers the assembly of ATP at the expense of the exchange of intracellular potassium for extracellular sodium.

The ionic inequalities that are maintained between living cells and their surroundings by the operation of the sodium-potassium pump are of crucial functional importance. They provide the cells with an internal milieu adapted to the requirements of their enzymes. In given cell types, they may drive other active-transport mechanisms, such as an antiport expulsion of calcium ions or a symport uptake of glucose or amino acids, coupled with

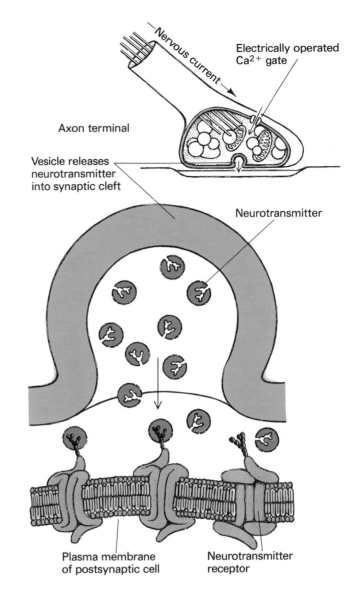

Schematic representation of synaptic transmission.

the exergonic re-entry of sodium ions into the cells. They are also responsible for the existence of a difference in electric potential at the cell surface. This resting potential usually lies between 50 and 100 mV and is created largely by the outward leakage of potassium ions, which explains why the ΔG of potassium transport is essentially zero (Appendix 2): net transport stops when the electric potential generated by outward diffusion is just sufficient to drive electrochemical transport at an equivalent rate in the opposite direction. Finally, the ionic inequalities produced directly and indirectly by the sodium-potassium pump have the advantage that they lend themselves to very rapid perturbations. Innumerable phenomena, including all the major operations carried out by the nervous system, depend on such perturbations, which are themselves elicited by means of regulated channels (gates).

Like permeases, gates allow only passive transport down concentration gradients. But instead of carrying their substrates across by means of binding sites, they let them slip through some sort of inner conduit, or channel. They do have binding sites, but these serve to control the opening or closing of the channel by regulating ligands. Gates also exist that respond to changes in certain ionic gradients or in membrane potential. The subtle interplay of chemical, ionic, and electrical factors whereby these various gates are controlled underlies much of the functioning of the nervous system, including the manner in which the ten-billion-odd neurons that make up the human brain communicate with each other and with other cells in the body to stage the most prodigious manifestations the biosphere has to offer.

Nerve Transmission

Only a detailed and extensive tour of the nervous system could do justice to even the little that is currently understood of its stupendous complexity. A fleeting visit is all we can afford; it is enough, however, to give us an inkling of what seems to be a basic functional blueprint. For this purpose, let us follow, at least in thought (we would need a fast car to do it in reality), the wave of depolarization that travels down the axon of a motor neuron that has just been stimulated. As the nervous current reaches one of the small knobby bulges whereby the axon connects with a muscle cell at the neuromuscular junction, it causes the opening of an electrically operated calcium gate. Calcium ions rush in and induce the exocytic discharge of small membrane-bounded vesicles, about 40 nm in diameter, which occupy this region of the cytoplasm. Each of these synaptic vesicles, as they are called, contains some ten thousand molecules of acetylcholine, a substance that we encountered in Chapter 8 as the product of a three-step, CoA-mediated, β_p type of assembly between acetic acid and choline.

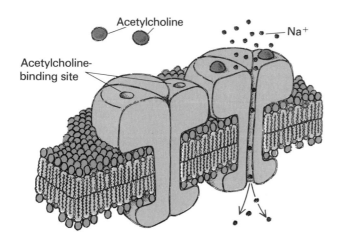

Acetylcholine

Acetylcholine-
binding site

Na⁺

Model of acetylcholine receptor. Two identical protein subunits bearing an acetylcholine-binding site join with three other subunits (only one is shown) around a central channel, which is closed when the binding sites are unoccupied. The binding of acetylcholine (green) causes the channel to open and let sodium ions (red) through.

When released extracellularly, acetylcholine survives less than a thousandth of a second because of the presence of a powerful enzyme, acetylcholinesterase, that immediately hydrolyzes it. But this brief time is enough for significant amounts of the substance to diffuse across the narrow space (synaptic cleft) that separates the two cells and to occupy specific binding sites on the surface of the muscle cell. These binding sites belong to a protein complex, the acetylcholine receptor, which is a ligand-operated sodium gate. When the receptor is occupied by its specific ligand, the gate opens and admits sodium ions into the muscle cell. The resulting depolarization of the cell membrane is propagated by neighboring electrically operated sodium gates and, through the T-system (Chapter 12), activates another electrically operated gate, which lets calcium out of the sarcoplasmic reticulum. And so we are back to calcium, this time as trigger of actin-myosin interaction: the affected myofibril contracts.

All these events are very short and finely tuned. We have already seen the role of acetylcholinesterase in the removal of acetylcholine. Pumps, on the other hand, quickly correct the ionic perturbations. As to the synaptic vesicles, they are replaced, though more slowly, by the new formation of acetylcholine and the endocytic recovery of membrane material. The scene is soon reset for a new act.

This script, with a variety of different settings and protagonists, tells the story of virtually every communication between a nerve cell and another cell, be it another neuron, a muscle cell, a secretory cell, or some other effector cell. Except for those rare cases where electric impulses travel directly between cells, such as through gap junctions (Chapter 2), the signal is always conveyed chemically, by means of a neurotransmitter discharged by the excitatory nerve cell and acting on the target cell through the mediation of specific surface receptors. Many different neurotransmitters besides acetylcholine are known. They include norepinephrine, epinephrine, histamine, serotonin, dopamine, glycine, γ-aminobutyric acid (GABA), and a variety of peptides. The corresponding neurons are designated cholinergic, adrenergic, histaminergic, and so forth. The total number of neurotransmitters is unknown but could be well in excess of one hundred. They are particularly numerous and important in the brain, where they provide the elements of an elaborate chemical communication network.

As there are many transmitters, there are also many receptors—more than there are transmitters, in fact, as some transmitters, such as acetylcholine itself or epinephrine, act on more than one type of receptor. Some of these receptors control ionic gates, as does the acetylcholine receptor, but with a variety of effects, depending on what ions are let in or out, and where. Other receptors act differently, as we are about to see.

The extent of our dependence on neurotransmitters is truly staggering. The regular beating of our hearts, the maintenance of adequate blood flow and pressure in our arteries and veins, the continuous monitoring of blood gases, with immediate correction of any disturbance by appropriate changes in the depth and rhythm of our breathing, the whole hidden orchestration of peristaltic propulsions and digestive secretions needed for the processing of our meals are just a few among the countless silent coordinating activities that rule the functioning of our organs with the help of neurotransmitters. And this is only part of the story. Every one of our movements, voluntary or involuntary; all of the information we receive from the outside world through our sense organs; the sum total of our thoughts, impulses, emotions, and dreams, together with their substratum of unconscious and subconscious data-processing and integration—all rely critically on appropriate neurotransmitter molecules being released at the right time and in the right amounts at billions of different intercellular connections; on their interacting correctly with their receptors to produce their effects; and on their being removed or destroyed at exactly the rate needed for the effects to last long enough, but not too long.

Not surprisingly, a network of this sort is exposed to an immense variety of interferences. For each authentic neurotransmitter, hundreds of different imperfect copies (analogues) can be made that retain the ability to interact with the receptor and either mimic (false keys) or block (jamming devices) the effect of the natural transmitter. Furthermore, release of the real transmitter may be induced, enhanced, or inhibited; its synthesis or breakdown may be accelerated or slowed down. This explains the wealth of substances, both natural and synthetic, that affect the nervous system. From the deadly paralysis spread by a curare-dipped arrow to the horrible convulsions induced by extracts of the strychnos nut, from the cosmic serenity of the opium smoker to the piercing hallucinations of the LSD victim, from pep pills to tranquilizers, from knock-out drops to nerve gas, the whole gamut of neuropsychotropic drugs and poisons created by the combined ingenuity of Man and Nature is nothing but a huge collection of perturbing agents capable of interfering, in an infinite variety of ways, with chemical neurotransmission.

If so much can be done from the outside to our thoughts and moods, why not also from the inside? This question is being asked with increasing insistence by modern neuropsychiatric investigators who are searching very hard to give precise chemical identities to the demons of the past and to their more recent Freudian substitutes. Some may deplore this attempt at demythification of mental illness, but not so the patients the day they find that a simple medication can save them from protracted exorcisms on a psychoanalyst's couch or from commitment for life to the company of their fellow sufferers.

Hormonal Effects

Many receptors have more than a simple gating function and are connected to more or less complex enzyme systems that face the cytosol on the *cis* side of the plasma membrane. Best known of these systems is adenylate cyclase, an enzyme that catalyzes the formation of cyclic AMP from ATP (Chapter 8):

$$ATP \longrightarrow cAMP + PP_i \ (2 \ P_i)$$

The product of this reaction is a multipurpose chemical messenger that is dispatched from its receptor-linked site of formation on the plasma membrane to various intracellular systems that are sensitive to it, in particular certain protein kinases. The important regulatory function fulfilled by this class of enzymes was mentioned in Chapter 12 in a discussion of the biological effects of calcium ions. A number of hormones, including some that double as neurotransmitters (e.g., epinephrine), act via cAMP. The selectivity of the effects of each hormone is ensured by the localization of its receptor. For example, ACTH, the pituitary adrenocorticotropic hormone, switches on cAMP-dependent steroid formation in the adrenal cortex without activating cAMP-dependent mechanisms elsewhere because its receptors occur predominantly on the surface of adrenocortical cells. Conversely, the effects of parathyroid hormone, which also stimulates cAMP production, are restricted to certain cells in the bones and kidneys that display the appropriate receptor molecules.

A remarkable characteristic of these mechanisms is their power of amplification. To appreciate this, just take a look at a liver cell that is being stimulated by epinephrine or glucagon to fragment glycogen and produce glu-

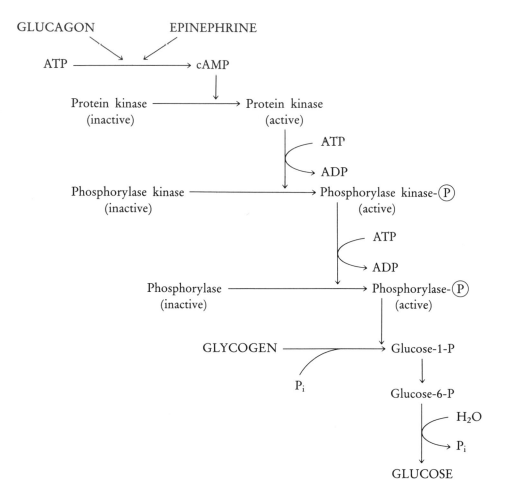

cose. Binding of either hormone to its receptor triggers the production of cAMP, which stimulates a protein kinase to activate, by phosphorylation, another protein kinase, which itself phosphorylates an enzyme called glycogen phosphorylase, which, in phosphorylated form, catalyzes the breakdown of glycogen to glucose-1-phosphate, a precursor of glucose, as shown above.

Almost every step in this cascade is catalytic. Let us assume conservatively that each catalytic step causes a hundredfold amplification: then one molecule of hormone stimulates the production of 100 molecules of cAMP; each molecule of cAMP-activated protein kinase phosphorylates 100 molecules of phosphorylase kinase; and so on. The end result is the production of 100 million (10^8) molecules of glucose for every molecule of hormone bound to its receptor. Thus, minute amounts of hormone can have dramatic effects. Note that there is a price to pay for this exquisite sensitivity: 10,101 ATP molecules (one to make cAMP, plus 100 to phosphorylate phosphorylase kinase, plus 10,000 to phosphorylate phosphorylase). This price is really negligible, however, in regard to the 38 molecules of ATP the cell can gain from the oxidation of each molecule of glucose, which means that the cell spends one ATP to get back 380,000. It is a bargain.

A more serious drawback of high amplification is the danger of things getting out of hand. It is minimized by the occurrence of numerous safeguards, including interiorization and destruction of the hormone, participation of antagonistic hormones, inactivation of the receptor, hydrolysis of cAMP to AMP by a specific phosphodiesterase, dephosphorylation of phosphorylase or of its kinases by phosphatases. But when these safeguards fail or are inoperative, tragedy may ensue. This happens in cholera. The microbe responsible for this terrible disease secretes a toxin that binds in a stable fashion to an adenylatecyclase–linked receptor present on the surface of intestinal cells. These are thereby flooded continuously with high amounts of cAMP, which, among other effects, stimulates the cells to excrete salts and water uninterruptedly. Hence the persistent diarrhea and severe dehydration that characterize the disease and frequently cause the death of its victims.

Mitogenic Stimulation

A particularly important, as well as mysterious, class of receptors comprises those that evoke a mitogenic response. When turned on by a ligand, they start a complex sequence of events that many hours later, long after the ligand has been shed from the receptor, ends up with DNA duplication, construction of a mitotic apparatus, and cell division. The control of cell multiplication by trophic hormones or growth factors, crucially important for the generation and maintenance of each cell population in the organism, takes place through the mediation of such receptors. Somewhere along the chain of events that they control lie points through which cancer-producing agents exert their deregulating effects. There is evidence that certain special protein kinases play an important role in both normal and abnormal mitogenesis (see Chapter 18).

Mitogenic control is developed to an astounding degree of diversity and selectivity in the lymphocytes. As a result of a unique process of ontogenic diversification (see Chapter 18), the issue of this line of cells consists of literally hundreds of millions of distinct individual types, each of which has its mitogenic signaling device hooked to a different surface receptor. The number and variety of these receptors is such that virtually every possible macromolecule—except those of the organism itself, for which immunological tolerance has been built up during fetal life—is recognized and bound by at least one subclass of lymphocytes. The cells to which this happens are thereby led to multiply and to form a clone. The ligands that trigger such a mitogenic response are called antigens. In the B-lymphocyte line, the cells involved are those that manufacture an antibody directed against the stimulating antigen. (The receptor is none other than the antibody itself, anchored on the cell surface by an extra hydrophobic tail.) In T lymphocytes, the cells generated by mitogenic stimulation are killer cells that selectively destroy any cell that bears on its surface the antigenic groupings that started the multiplication process. This is how the immune system builds up specific army corps against each individual challenge by a foreign molecule or organism (Chapter 3). Certain glycoproteins, called lectins, mostly of vegetable origin, have the ability to stimulate lymphocyte proliferation nonselectively. Examples are phytohemagglutinin (extracted from kidney beans) and concanavalin A (from jack beans).

Endocytic Uptake

A general function common to many receptors is to mediate the selective endocytic uptake of their ligands, usually for subsequent delivery to the lysosomes and consequent degradation. As we saw in Chapters 4 and 5, all sorts of important regulatory mechanisms have been built around this primitive feeding process. In the framework of receptor function, it has a number of important aspects. First, there are cases where uptake conditions the cell's response to a bound ligand because the triggering change in the occupied receptor takes place only in an acidic milieu, such as is provided in endosomes and lysosomes. Epidermal growth factor (EGF), a mitogenic hormone acting on epithelial and fibroblastic cells, is said to have such a requirement. So does the invasion of the cell by a number of membrane-wrapped viruses, as we can testify from personal experience (Chapter 7), as well as the intracellular penetration of various protein toxins (see p. 234). Another consequence of endocytic uptake is to limit the duration and intensity of a response by bringing about the destruction of the ligand. As to the receptors themselves, they may resurface, ready to function again, as early as a few minutes after interiorization, thanks to the phenomenon of membrane recycling (Chapter 6). Or, as happens with some hormone receptors, they may accompany their ligands into the lysosomes and be inactivated, with the consequence that the cells become less responsive to the hormone (down regulation).

A common characteristic of endocytic receptors is some form of connection on the *cis* side of the plasma membrane with an elaborate cytoskeletal rig that does the mechanical tasks of endocytic uptake. Clathrin baskets (Chapter 12) are the best-known such devices. But there may be others, in which actin fibers do the pulling, probably in combination with myosin. To what extent occupancy of the receptors conditions uptake is not clear and may vary according to receptor (Chapter 4). The same is true of the clustering phenomenon that causes the receptors to congregate on the membrane patches that are pulled in. Some receptors naturally home to coated pits, perhaps because they are linked on the *cis* face to clathrin or to some clathrin-associated protein. Others are brought together by divalent or multivalent ligands on the *trans* face. The most dramatic such event is capping, a phenomenon in which all the occupied molecules of a given receptor form a single aggregate, which is then interiorized in one big endocytic gulp.

Chemotaxis

Possibly related to certain forms of receptor-mediated endocytosis is the phenomenon of chemotaxis. (Remember, it was one of the very first sights to meet our eyes when we set off on our tour.) The receptors involved in this remarkable process are linked dynamically to the cell's locomotor system in such a way that its response does not depend simply on the number of occupied receptors, but also on their spatial distribution. This distribution will be uneven whenever the cell is exposed to a flux of ligand molecules diffusing down a concentration gradient: there will be a corresponding gradient of occupied receptors on the cell surface, reflected inside the cell by a similar gradient of whatever messenger molecules or other activating devices are triggered by the occupied receptors. Such a gradient seems to be all that is needed to orient the complex mechanisms whereby filopodia are pushed out, adhesion plaques attached, and actin cables retracted to move the cell toward or away from the incoming ligand molecules. Ligands capable of influencing cell movement in this way are said to be positively or negatively chemotactic. They act as cellular attractants or repellents.

Translocation

One of the strangest forms of exploitation of surface receptors is exhibited by certain toxic proteins manufactured by bacteria (e.g., diphtheria toxin) or by plants (e.g., ricin, extracted from castor beans). These toxins consist of two parts linked by disulfide bridges: the A chain, or "effectomer," is an enzyme with the property of rendering inactive some component of the protein-synthesizing machinery; the B chain, or "haptomer" (Greek *haptein*, to bind), is a ligand with the ability to bind specifically to certain surface receptors. By itself, the effectomer is absolutely harmless because it cannot get through the plasma membrane. But, when it is brought close to the membrane by its haptomer carrier, it gets detached from the haptomer and translocated into the cytoplasm, where it does its deadly work. In several such cases, the actual piece of legerdemain takes place only after endocytic uptake and is triggered by the endosomal acidity.

This phenomenon has raised questions regarding its possible physiological significance. Several hormones resemble the toxins. In particular, the pituitary thyroid-stimulating (TSH), follicle-stimulating (FSH), and luteinizing (LH) hormones, as well as the placental choriogonadotrophin, are built of two pieces. One, common to all four, is an adenylate cyclase activator. The other carries the tissue specificity of the hormone and serves as a homing device by recognizing receptors that are uniquely present on the surface of the hormone's target cells. The analogy with the effectomer-haptomer combination of the toxins is suggestive. There is, however, no evidence that the action of these hormones requires their endo-

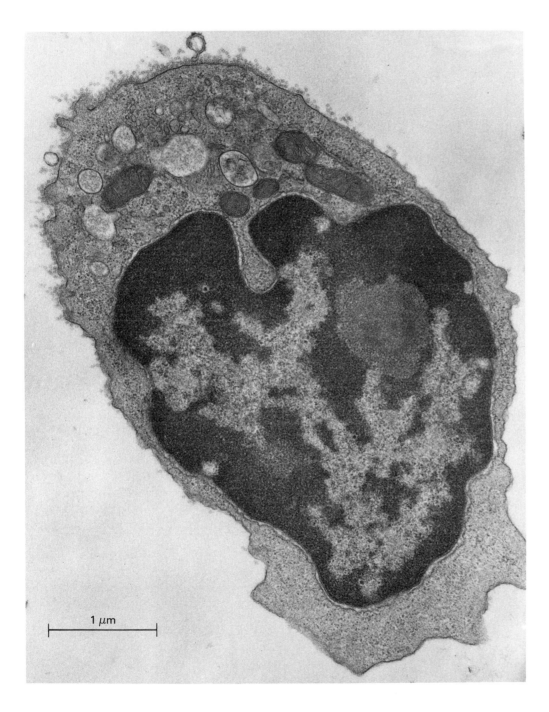

Capping. A mouse B lymphocyte, which normally has immunoglobulin receptor molecules distributed over its whole surface, has been treated with an antibody against this receptor, resulting in the clustering of all the receptor molecules at one pole (upper left) of the cell. This "cap" is made evident by the covering of small square or round markers, which are molecules of the giant protein hemocyanin (the blue copper-containing blood protein that serves as oxygen carrier in the snail) that have been attached chemically to the antibody molecules.

1 μm

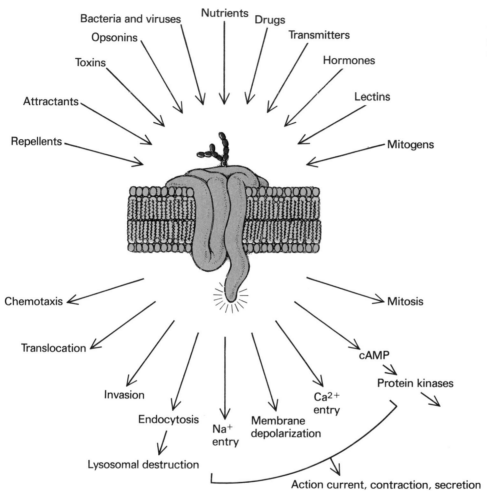

Bacteria and viruses
Nutrients
Drugs
Opsonins
Transmitters
Toxins
Hormones
Attractants
Lectins
Repellents
Mitogens

Chemotaxis
Mitosis
Translocation
cAMP
Invasion
Protein kinases
Endocytosis
Ca^{2+} entry
Na^+ entry
Membrane depolarization
Lysosomal destruction
Action current, contraction, secretion

Some channels of communication established by surface receptors between the outside and the inside of the cell.

cytic uptake and introduction into an acidic milieu or involves passage of the effectomer part across the cell membrane.

So goes the dance of the receptors, a fantastic display of virtuoso acts staged on the inner face of the plasma membrane by all sorts of transducers and signaling devices and choreographed on its outer face by hosts of transmitters, hormones, mitogens, lectins, nutrients, ions, opsonins, carrier proteins, attractants, repellents, toxins, drugs, and even microorganisms and viruses, all sharing the key property of being able to bind specifically to some surface receptor. The ligands call the dance, the receptor-linked transducers command the steps, but only those cells respond that own the particular receptor capable of connecting the ligand to the transducer. Therein lies the specificity of innumerable biological responses, interactions, and effects.

A necessary corollary of this general mechanism is that each type of cell has its own combination of surface receptors. Cells, therefore, betray their identity—and thereby lay themselves open to all sorts of outside manipulations, even selective killing—to any outsider cognizant of the receptor code. No wonder, then, that therapy-motivated cell explorers who dream of "guided missiles" and other "magic bullets" as means of subduing recalcitrant cell populations expend so much time and effort scrutinizing the hundreds of different receptors exposed on the surface of the cells. It is one of the hot spots of contemporary cell research.

The Cytosolic Face of the Vacuome

In line with their role as boundaries of spaces that, directly or indirectly, make transient connections with the outside world and with their participation in the vesicular carousel that moves material in bulk into, through, and out of cells, the membranes of the vacuome share many of the properties of the plasma membrane. They bear pumps, permeases, and transport receptors and are wired to cytoskeletal elements that direct their movements. This is particularly true of the exoplasmic parts (Chapter 6), many of which are continuously recycling between the interior and the surface of the cells.

We came across some of these features in our wanderings. We saw a proton pump, possibly imported from the plasma membrane, working to acidify endosomes and lysosomes (Chapters 4 and 5) and learned of the probable participation of permeases in the clearance of lysosomes (Chapter 5). We found out in Chapter 12 that striated muscle cells subdue their myofibrils with the help of a calcium pump—a device that operates on a phosphorylation-dephosphorylation cycle, as does the sodium-potassium pump—situated in the sarcoplasmic reticulum. And, as we have just seen, they depend on an electrically controlled calcium gate inserted in the same membranes to call the myofibrils to action. In Chapter 6, the general significance of receptor clustering on movable membrane patches was noted as a mechanism of selective sorting and transport, exemplified by the mannose-6-phosphate receptor for lysosomal enzymes. Also discussed were some of the logistics of transport mechanisms that rely on an accurate mutual recognition and intimate apposition between two membranes and on the resulting fusion and reorganization of their lipid bilayers, which nevertheless occur without significant intermingling of their protein components.

Biosynthesis in the ER

Besides such general features common to all these membranes, there are important differences. Just as we contemplated many different kinds of surfaces when we toured the vacuome, so now our cruise along the cytosolic boundaries of this membranous system presents us with a similar number of different sights. Endoplasmic regions, in particular, show conspicuous signs of an intense biosynthetic activity, of which we have seen only the end result so far (Chapter 6). Most impressive is the assembly of proteins by the polysomes that cover the rough-surfaced parts of the ER. All of Chapter 15 deals with this process, and so we will skip the ribosomes for the time being. But there is much more to be seen and enjoyed, especially by those who followed our biosynthetic excursion in the second half of Chapter 8.

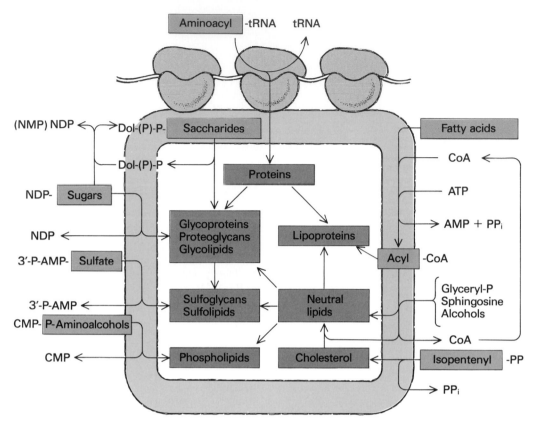

Summary of the main biosynthetic processes that take place on ER (and Golgi) membranes. Note that, with the exception of fatty acids, which are activated locally, all X-O⁻ building blocks (shown in blue, Chapter 8) are delivered in activated form by carriers to the assembly sites. Biosynthetic products accumulate in the luminal space or, as is the case for some proteins and many lipids, remain membrane bound.

Between the attached ribosomes, a variety of nucleoside diphosphates (Chapter 8) deliver activated glycosyl groups for the assembly of the oligosaccharide side chains of glycoproteins. These chains are stubby, bushlike structures containing from eleven to fourteen sugar molecules. They are assembled on a carrier, dolichyl diphosphate, which is firmly rooted in the lipid bilayer by a long hydrophobic tail. As we saw when we were inside the ER (Chapter 6), completed shrubs are transferred en bloc onto growing polypeptide chains emerging in the vicinity. The whole process is quite baffling, as it requires the translocation, sometimes with the help of other dolichyl phosphate molecules, of the highly hydrophilic glycosyl groups across the lipid bilayer. Sulfation and phosphorylation of some of the carbohydrate side chains also take place through the membrane from carriers charged in the cytosol.

Other areas of the ER membranes are occupied with lipid biosynthesis. Some of the more hydrophilic building blocks needed for this process—for instance, glycerol,

choline, or ethanolamine—are brought by special nucleotide carriers (Chapter 8) from cytosolic activating sites. Others, especially the fatty acids, are activated locally and linked to coenzyme A, from which they are transferred to their final acceptor. These are busy factories making all sorts of neutral fats and phospholipids, as well as cholesterol and its derivatives, and many other lipid-soluble molecules. Sometimes the lipid and carbohydrate factories converge to make glycolipids. Some of these lipid products are transferred into the lumen of the ER, where they are combined with proteins to form secretory lipoproteins. Others are stored in the cytosol, where they aggregate into oily droplets, immiscible with water. Yet others remain in the membranes, as do some of the newly made proteins, thus ensuring the manufacture of new membrane material.

Conjugation

Not directly related to biosynthesis as a self-building activity, but relying on the same type of mechanism, are the various phenomena known as conjugation. They can affect a variety of lipid-soluble molecules, some of them made physiologically (e.g., cholic acid), others introduced from the outside (e.g., a number of drugs); typically, they link these molecules with some hydrophilic substance such as glucuronate, sulfate, acetate, or some amino acids. The main consequence of conjugation is that it renders the affected molecules more water-soluble and favors their excretion in bile (bile acids) or urine. Most such processes occur on the membranes of the ER and use carrier-linked activated building blocks, as do the analogous authentic biosynthetic reactions.

Hydroxylation

One would think that such manifold group-transferring activities would suffice to occupy all the available sites on the membranes. But this is hardly so. Actually, a careful look will show that ER membranes appear as a quiltwork of white and pink patches and that unloading of activated groups occurs mostly on white patches. What, then, is the function of the pink patches? Color, in the biological world, is often associated with electron transport. The pink areas of the ER are no exception. They are occupied by two cytochromes, known as b_5 and P_{450}, and are involved in a very special type of electron transfer, from NADPH to molecular oxygen, that takes place in such a way that half the oxygen is reduced to water and the other half serves to hydroxylate some organic substrate:

$$NADPH + H^+ + O_2 + R\text{—}H \longrightarrow$$
$$NADP^+ + H_2O + R\text{—}OH$$

This is one of the rare reactions in which oxygen is not used simply as electron acceptor but actually becomes attached to a metabolite.

A remarkable property of this system is its lack of specificity. It has its physiological substrates, the most important ones being steroid hormones and their precursors. But it will also act on innumerable artificial chemicals, including medicinal drugs and environmental pollutants. In many instances, the change inflicted on the substrate decreases its pharmacological or toxic properties

and favors its elimination, either as such or conjugated. For this reason, the hydroxylating system of the pink ER membrane patches has often been referred to as a detoxicating system, and its lack of specificity regarded as a benevolent gift of natural selection, which somehow, millions of years ago, prepared our cells for the man-made chemical invasion of the world. But now we must modify this optimistic view, for it also happens that harmless molecules are rendered toxic by the action of the hydroxylating system. The kind of toxicity it conveys is a particularly treacherous one, as it often includes the ability to cause mutations and cancerous transformations. Many environmental carcinogens do not possess, as such, the property to cause cancer; they acquire it when they come into contact with the ER.

Except for the polysomes and their anchoring proteins (ribophorins), which are concentrated in the more remote parts of the ER, all these systems seem to be distributed throughout the whole membrane network, presumably because they are free to diffuse laterally along the lipid bilayer and thereby tend to occupy in uniform fashion the whole area available to them, forming only such microclusters as arise from their mutual interactions. This is indeed what we observe until we reach the junction between the smooth ER and the Golgi apparatus. There the scene changes abruptly. Enzymes characteristic of the ER disappear and are replaced by others. For instance, we find in the Golgi membranes a new family of glycosyl transferases that receive their activated sugar molecules from the same NDP carriers as do those located in the ER but transfer them to their final acceptors without the help of dolichyl phosphates. These enzymes finish off the job started in the ER. In addition to constructing new chains, they remodel those built in the ER, add new sugar molecules to them, often after their partial dismantlement by glycosidases, and cap them with special molecules (fucose, sialic acid) or groupings, including the mannose-6-phosphate tag of lysosomal enzymes. They also play an important role in fitting the membrane proteins themselves with carbohydrate side chains.

These processes are consistent with the position of the Golgi apparatus as a halfway station between the ER and the plasma membrane. But how changes in membrane composition are generated and maintained, in spite of the extensive opportunities for intermingling offered by the multiple associations between the Golgi and other types of membranes, remains a very puzzling problem.

The Mitochondrial Boundary

Like their bacterial counterparts, the energy-transducing membranes of mitochondria are authentic boundaries that mediate an intense, strictly regulated molecular traffic. The outer mitochondrial membrane has little to do with this regulation. It is an essentially open frontier of somewhat uncertain function. Virtually all the control is exerted by the inner membrane. This is understandable. To function as chemiosmotic transducer, the mitochondrial inner membrane must be tightly sealed. Hence, it alone can control such channels as have to exist to allow the cell's power plants to operate. This adds even more to the intricacy of its architecture, interspersing a network of molecular checkpoints between the protonmotive, electronic microchips that delighted us with their astonishing construction when we first set eyes on them. Most mitochondrial transport systems are of the passive antiport kind, but some are linked to protonmotive power and can drive their substrates uphill. Mitochondria, like other power stations, keep up three types of exchanges with their environment. They import fuel, export power, and undergo repair and maintenance. They do, however, differ from man-made power stations by their greatly superior versatility and efficiency. As we have seen, the mitochondrial matrix contains systems capable of oxidizing all major foodstuffs. Accordingly, ports exist on the inner membrane to admit a number of organic acids, including pyruvic acid, amino acids, and fatty acids. The last do not enter as such, but in activated form, with the help of carnitine, a special carrier that mediates the transfer of fatty acyl groups from cytosolic to mitochondrial coenzyme A. Fuel entry is complemented by an equivalent influx of oxygen, probably facil-

itated by the oxygen-consuming cytochrome-oxidase complex.

Systems also exist that allow the fueling of mitochondria with high-energy electrons produced in the cytosol—for instance, by glycolysis. These electrons are ferried by special redox couples that collect them from NADH in the cytosol, deliver them inside the mitochondria, and return to the cytosol in oxidized form to pick up another electron pair:

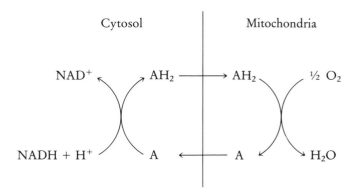

Several such systems are known. As shown by the diagram, they require two electron-transferring enzymes acting on the AH_2/A couple, one in the cytosol, the other in the mitochondria, as well as ports for the two members of the couple in the mitochondrial membrane. Their function is important because there are no passageways for NADH or NAD^+ on the mitochondrial surface. This is the only way in which cytosolic NADH can be reoxidized by oxygen.

Thanks to the existence of the Krebs cycle, mitochondria oxidize their fuel completely: they are nonpolluting power stations. The waste products flowing out of them are simple, nontoxic compounds, mostly carbon dioxide (as bicarbonate), water, inorganic sulfate, and, as sole potentially toxic substance, ammonia, which is tolerated by cells only at low concentration. In many animals and in humans, mitochondria are equipped with an antipollution mechanism that converts ammonia into urea. In birds and reptiles, ammonia is detoxified in the cytosol by conversion into uric acid.

In exchange for the fuel they receive, mitochondria export power in the form of the group potential of the γ bond of ATP assembled from incoming ADP and inorganic phosphate. A special ATP/ADP translocase allows exit of an ATP molecule in exchange for an entering ADP (or AMP, thanks to the occurrence of an AMP kinase in the space between the two membranes). Inorganic phosphate enters by an antiport in exchange for hydroxyl ions.

As already mentioned, protonmotive power can serve to move ions across the mitochondrial boundary, either by a symport or an antiport mechanism (e.g., inorganic phosphate) or by means of an appropriate ionophore or carrier with the help of the membrane potential (e.g., calcium). The key role played in many cells by the mitochondrial calcium pump was alluded to in Chapter 12. What is much more obscure is the manner in which mitochondria stock up on the materials needed to support the growth process, which, complemented by division, is the only mechanism whereby their populations are maintained in stationary cells and augmented in growing cells. Mitochondria can make many of their constituents, even proteins, from small building blocks. But, as mentioned earlier, they carry the genetic information for only a very small number of proteins. The instructions for most mitochondrial proteins are stored in the nucleus and translated in the cytosol by soluble polysomes. The synthesized proteins somehow work their way through the outer membrane and, in most cases, through the inner membrane, in spite of its hermetically sealed boundary. Most chloroplast proteins face a similar problem. So do those of peroxisomes, which also are made in the cytosol. The possible role of signal sequences, removable or not, in these posttranslational transfers will be discussed in Chapter 15.

14 | The Nexus of Metabolism: Integration and Regulation

During much of our visit so far, bioenergetics has provided the main framework within which to rationalize what we saw. There were good reasons for that. Thermodynamics is an essential and trustworthy guide. It distinguishes clearly what can happen from what cannot and helps us understand why things can happen the way they do. Especially when it comes to life and its astounding combinations of apparent improbabilities, the insights given by thermodynamics have proved most illuminating. But there is an important limit. Thermodynamics tells us what *can* happen; it does not tell us what *does* happen. For example, there is nothing from the energetic point of view to prevent the whole of the biosphere going up in flames. Why it does not, and why it sometimes starts doing so, as in a forest fire, is not a thermodynamic problem but a kinetic one. This, at least, is how we distinguish them on an elementary level. At a deeper level, the two aspects fuse within the framework of statistical thermodynamics.

The Kinetic Factor

Kinetics deals with rates. It helps distinguish, among the infinite number of events that are thermodynamically possible, those few that occur in reality at measurable rates. This distinction is doubly important for living cells: first, in something of an all-or-none fashion, because most of the events that take place in living cells are of the kind that, although allowed by thermodynamics, do not take place to any significant extent without a suitable catalyst. What a cell can do, therefore, is determined by its enzymic equipment, itself a reflection of the cell's genetic endowment. In this respect, much of the diversity within the biosphere boils down to the ability, or inability, of organisms to make specific enzymes. This is true even within a given species. In humans, many congenital abnormalities result from severe deficiencies of single enzymes.

But it is not enough for a cell or organism to be able to carry out a given reaction or process. The rate at which it does the job is often crucially important as well. Regulation and adaptation are accomplished to a large extent by rate modifications. These sometimes have themselves the character of an all-or-none process. A number of the switches operated by surface receptors are of this kind, at least in first approximation. More often, the response is graded and is both mediated and modulated by more or less intricate feedback loops. We cannot leave the cytosol without taking a brief look at these mechanisms. Actually, we can see them in action everywhere in the cell. But there is no better place than the cytosol, which, in addition to being a major cell organ in its own right, containing hundreds of enzymes and up to half a cell's total protein, represents a central communication system, a unifying connecting link, the obligatory intermediate in all the exchanges and interactions that take place between the different cellular organelles.

The Simple Rules of Molecular Circulation

Much of the complex molecular traffic that fills the busy cytosolic thoroughfares is readily understood with the help of the simple law of diffusion. Soluble molecules, with no source of energy other than random thermal agitation, move spontaneously from any point where their concentration is higher to one where it is lower—down their concentration gradient. This is the direction imposed by thermodynamics (Appendix 2). Kinetically, however, diffusion is a very slow process, which becomes rapid only in situations where large concentration differences are maintained over extremely short distances—that is, where the concentration gradient is very steep. Such is the case in living cells. Indeed, once aware of the general properties of the diffusion process, we have no difficulty recognizing in the cytosol countless molecular fluxes guided by concentration gradients, often of highly complex three-dimensional geometry, and leading invariably from sites where substances are generated or admitted to areas where they are altered or let out. Some of these fluxes have been mentioned before, and the concept may now be generalized.

There are, however, cases where diffusion is too slow for the needs of the cells in spite of the short distances to be covered, because nowhere does the substance to be moved reach a high enough concentration. This may be so because of unfavorable equilibrium conditions, as is the case for some high-energy biosynthetic intermediates. Some of the devices whereby such thermodynamic hurdles are overcome have been alluded to in Chapter 8. Low solubility in water is another factor that may prevent a substance from building a steep enough concentration gradient. Most lipid-soluble substances share this drawback. Such substances are transported by carriers, generally of protein nature, which either move by simple diffusion themselves or form a chain across which the substances move by exchange, jumping from one carrier molecule to another (exchange-diffusion).

Oxygen is a special case. It enters cells by diffusion through whichever face is closest to the blood supply and then it automatically draws toward the mitochondria, each of which maintains an oxygen "sink" around itself. Fortunately, mitochondrial oxidations can function at maximal capacity even at very low oxygen concentration (owing to the high oxygen affinity of the respiratory enzyme cytochrome oxidase) and, thanks to this property, can usually generate a steep enough gradient to satisfy their oxygen requirements. An exception is the powerful, ultrafast, fuel-guzzling, red striated muscle cell, which is helped by a special pigment—myoglobin, a close relative of blood hemoglobin—that serves both to store oxygen intracellularly and to facilitate its diffusion.

In many instances, however, whether in muscle or

other cells, the limiting factor in providing oxygen to the mitochondria is not intracellular diffusion but extracellular supply from the lungs by way of the blood. If this supply is inadequate, the cell has no solution other than to switch over to anaerobic glycolysis if it is to satisfy its energy needs. During a maximal effort, such as a 100-meter run, an athlete's muscles need more ATP than can be supplied by oxidative phosphorylation with the oxygen that is made available to the mitochondria. The ATP deficit is made up by anaerobic glycolysis, which, as we shall see, is started automatically under such conditions. The runner goes into oxygen debt and accumulates lactic acid. During recovery, part of this lactic acid is oxidized, and the remainder is converted back into glycogen by reverse glycolysis with the help of the ATP generated by the oxidative process. The debt is paid back—at the cost, of course, of a small entropic levy caused by the extra run, first down and then up again, along the glycolytic chain.

Incidentally, art lovers will derive a rare form of delectation from going through a cell that is experiencing fluctuations in oxygen supply. For understandable reasons, when the oxygen tension falls, electrons jam up the respiratory chain, which means that a larger proportion of each electron carrier in the chain will be in reduced form and a smaller one in oxidized form. The opposite takes place when oxygen again becomes more abundant. Now it so happens that the various flavins, cytochromes, and other colored members of the respiratory chain have different colors or shades in oxidized and in reduced form. Thus, with fluctuating oxygen tensions, the mitochondria go through subtle changes in hue most pleasing to behold. The observer equipped with special spectroscopic lenses, especially if they reach into the ultraviolet region, will derive even greater enjoyment, as well as much detailed information on where and to what extent electron flow is hindered. Both the effects of oxygen tension and those of numerous inhibitors can be gauged and pinpointed in this way—and have been extensively by the adepts of this kind of spectrophotometric exploration.

An important property of the molecular fluxes that pervade the cytosol is that they are self-regulating. This is because the physical or chemical reactions whereby sub-

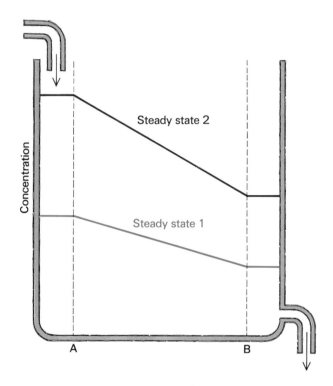

Schematic diagram of automatic adjustment from one steady state to another.

Steady state 1: The production of the substance in A, its diffusion from A to B, and its consumption in B occur at the same rate; the concentration profile remains unchanged.

Steady state 2: After the rate of production in A doubles, the concentration profile changes until the rates of diffusion and consumption double.

stances are caused to disappear tend to proceed faster when the concentration of their substrates increases (law of mass action). Take the simple example of a substance that is generated at site A and consumed at site B. Starting from a situation where generation, transport, and consumption occur at the same rate, imagine a sudden increase in generation rate. It is easy to see how this will cause the substance's concentration to rise in A, with, as consequence, a progressive increase in the steepness of its concentration gradient and a resulting acceleration of its rate of diffusion from A to B, leading in turn to a rise of its concentration at B and, thereby, to an increase in its rate of consumption. These readjustments will continue until new concentrations are established in A and B such that the rates of diffusion and of consumption equal the new generation rate. A situation where fluxes have been equalized in this way is called a steady state. What our example illustrates is the inherent tendency of the system to adopt a steady state and to respond automatically to a disturbance with the establishment of a new steady state. This kind of consideration may be extended to the whole complex network of metabolic processes. It goes far in explaining the homeostatic properties of living cells, which in Greek means their ability to stay alike (*homoios*). But a complication arises from the fact that most metabolic reactions are catalyzed by enzymes.

Enzyme Rules

Enzymes, like receptors, operate on the lock-and-key principle. They have specific binding sites for their substrates, just as receptors have for their ligands. The difference is that substrates bound by enzymes undergo a chemical change mediated by what is known as the active, or catalytic, site, which is closely linked to the substrate-binding site(s). It is easy to see that the number of substrate molecules altered per unit of time (reaction rate) depends on: (1) the number of enzyme molecules available; (2) their activity, as measured by the maximum number of substrate molecules each enzyme molecule is capable of handling per unit of time; and (3) their degree of occupancy. Airlines deal with the same three variables. For instance, the number of passengers an airline transports across the Atlantic per month depends on: (1) the total number of seats available; (2) the time it takes an aircraft to do a crossing and be ready for another one; and (3) the occupancy ratio. It is clear that the first two factors are characteristics of the airline's fleet and set an absolute limit to the number of passengers that can be transported. Only the third one depends on the number of potential passengers—within the limits imposed by the first two, as many a victim of overbooking can attest from personal experience.

Similarly, the rate of an enzyme-catalyzed reaction is influenced by the substrate concentration within the limits set by the number and activity of the enzyme molecules present. Unlike airplanes, however, which generally

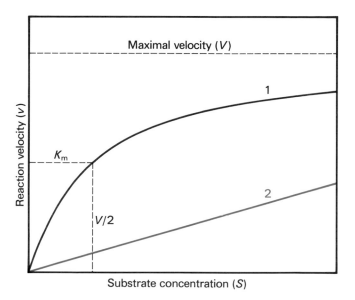

Enzyme kinetics: (1) hyperbolic relationship between the velocity of an enzyme-catalyzed reaction and the substrate concentration, according to the Michaelis-Menten equation; (2) linear relationship between the velocity of an uncatalyzed reaction and the substrate concentration, according to the mass-action law.

take up passengers as they come until the last seat is filled, enzymes fill their binding sites on a statistical basis. It may take tens or thousands or millions of substrate molecules dispersed in the cell volume to fill a single site, depending on the affinity between enzyme and substrate and on the number of sites remaining unoccupied. Mathematically, this relationship is expressed by the Michaelis-Menten law:

$$v = V \frac{S}{K_m + S}$$

in which v is the reaction velocity; V is the maximal velocity (the product of enzyme concentration by enzyme activity); S is the substrate concentration; and K_m is the dissociation constant of the enzyme-substrate complex (Michaelis constant)—that is, the inverse of the enzyme-substrate affinity constant (note that $v = V/2$ for $S = K_m$). Graphically, this equation is represented by a hyperbolic curve, in contrast to the straight line expressing the simple mass-action effect.

If we now refer back to the ability of living cells to readjust the rates of their metabolic reactions by simple mass-action effect, we find that the responsiveness of each individual reaction depends on the S/K_m ratio at which it operates; in other words, on how far the actual velocity v is removed from the maximal velocity V. For instance, if S equals one-hundredth of K_m (enzyme operating at about 1 per cent of its full capacity), the reaction velocity can be doubled by little more than doubling the substrate concentration. On the other hand, with $S = K_m$ (enzyme operating at half-maximal capacity), the reaction velocity cannot be doubled by any attainable increase in substrate concentration. Note that it can still be doubled by interventions at the enzyme level—for instance, by doubling the enzyme's catalytic activity or by doubling the number of enzyme molecules present (i.e., by doubling V). Most enzymes operate well below their maximal capacity, thus allowing ample scope for substrate-mediated, mass-action, regulating effects. The few that operate near maximal capacity catalyze the rate-limiting steps of metabolic pathways. They are the choice targets of enzyme-level regulatory influences.

The Three Levels of Metabolic Regulation

Within the substrate-mediated network, a particularly important role is played by the key coenzymes because of the large number of different reactions in which they participate. The most central role in this respect belongs to ATP and its hydrolysis products, ADP and P_i. They link the innumerable reactions that consume ATP to the oxphos units that regenerate it. Thanks to this link, the rate of catabolic electron flow, and thereby of foodstuff and acceptor consumption, is automatically adapted to the amount of work performed. We have already alluded to this property in reference to respiratory control in mitochondria. The diagrams on the facing page illustrate these relationships in highly schematic fashion for heterotrophs and autotrophs.

An essential feature of metabolic regulation is the existence of a loop whereby the products of anabolism can be used to fuel catabolism. The usefulness of this loop is evi-

dent. Without it, cells would have to balance their food intake and their energy expenditure from instant to instant, without any leeway—clearly a highly precarious situation, incompatible with the living conditions of most organisms. The existence of this important loop does, however, raise the question of why it does not operate continuously as a futile cycle, in which catabolism would endlessly undo the work of anabolism, with the useless squandering of energy as net result. Mythology abounds in such sad tales: Sisyphus's stone, the Danaids' pail, Penelope's tapestry. Were cells subjected only to mass-action types of regulating effects, nothing could prevent them from imitating these unfortunates.

Another problem concerns the choice of acceptor for the catabolic electrons. Oxygen has precedence, but why? This question was first raised more than a century ago when Pasteur discovered that the glucose consumption of yeast cells decreases dramatically in an anaerobic culture suddenly exposed to oxygen. To Pasteur, who was used to blowing dying embers back to life, this effect appeared paradoxical. The Pasteur effect has since been found to apply to many different types of cells; it has puzzled many investigators and inspired many hypotheses. Today we can account for most of it by simple homeostatic mechanisms. With the admission of oxygen, mitochondrial oxidations are set off and create an "electron sink" that draws (along the generated negative gradients) such available donors as pyruvate and the molecules capable of shuttling electrons from cytosolic NADH. Lactate is no longer produced. But, at the same time, the number of ATP molecules that the cell can obtain from each molecule of glucose metabolized increases nineteenfold. Therefore, to the extent that the cells do not perform more work or waste more energy in the presence of oxygen, a corresponding reduction in glucose consumption is to be expected from the ATP/ADP-mediated regulating mechanism that we have just examined. However, a problem remains. In glycolysis, the step that is controlled by the ATP/ADP ratio—the substrate-level oxphos unit—is preceded by two ATP-consuming reactions, which, as we saw in Chapter 7, are needed to convert glucose into phosphoglyceraldehyde, the actual substrate of the ox-

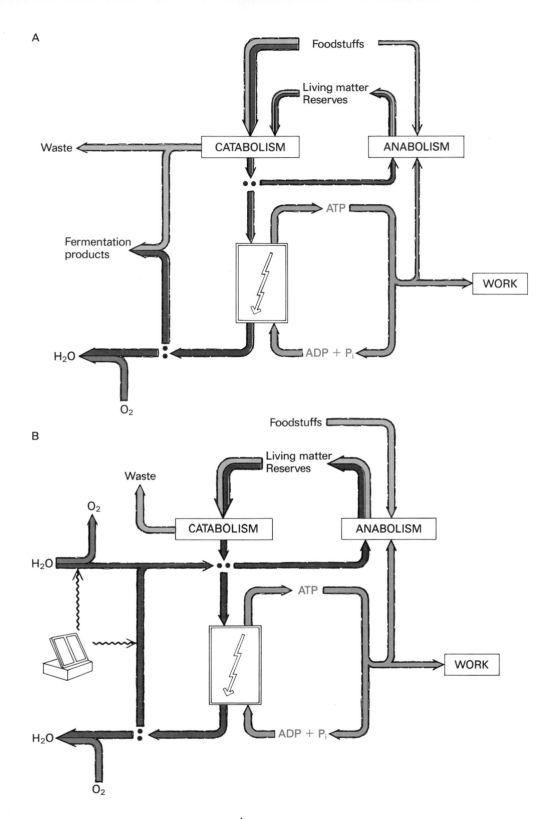

A

Foodstuffs

Living matter
Reserves

Waste

CATABOLISM

ANABOLISM

Fermentation
products

ATP

WORK

H_2O

ADP + P_i

O_2

B

Foodstuffs

Living matter
Reserves

Waste

O_2

CATABOLISM

ANABOLISM

H_2O

ATP

WORK

H_2O

ADP + P_i

O_2

The basic blueprint of metabolic regulation.

A. Heterotrophs. Foodstuffs from the outside serve mostly to support catabolism. The main role of catabolism is to detach high-potential electrons from the foodstuffs and to feed them into oxphos units. Exiting low-potential electrons are collected by an acceptor, either supplied separately (e.g., oxygen) or generated from the foodstuff flux (fermentation). The ATP assembled by oxphos operation supports the various forms of biological work. Side fluxes of foodstuffs, electrons, and ATP support the anabolic formation of biosynthetic products, including that of reserve substances. These in turn can be mobilized to support catabolism. Self-regulation of this system is achieved principally through the [ATP]/[ADP] [P_i] ratio, which automatically adapts the rate of catabolic breakdown to the amount of work performed (including anabolic work). If the work load increases, this ratio decreases, causing the rate of electron flow through oxphos units to increase through a mass-action effect. The opposite takes place if less work is performed. Fuelling needs are satisfied thanks to the links between catabolism and anabolism. When the supply of foodstuffs exceeds needs (feeding periods), excess substrates, electrons, and ATP converge to build biosynthetic stores. In the opposite situation (fasting, starvation), stores are drawn into catabolism.

B. Autotrophs (green plants). In the dark (delete light-powered photoreduction and photophosphorylation), the organism operates essentially like a starved heterotroph, subsisting on biosynthetic reserves and adjusting catabolism to work through the mediation of the [ATP]/[ADP] [P_i] ratio. In the presence of light, a supply of high-energy electrons supports all energy needs and electron-depleted, mineral foodstuffs can be used for anabolic purposes.

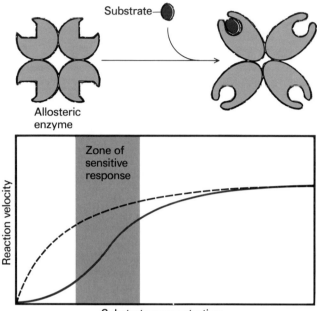

Reaction velocity

Zone of
sensitive
response

Substrate concentration

Allostery. An example of positive cooperativity. The binding of one molecule of a substrate to a tetrameric enzyme is hampered by the enzyme's low affinity for the substrate, but the binding of this first molecule causes a change in the conformation of the enzyme that greatly enhances its affinity for further substrate binding. The sigmoid shape of the kinetic curve illustrates this effect. Within a certain concentration zone, small changes in substrate concentration induce large changes in reaction velocity. Compare with normal hyperbolic curve (dashed line).

necks, ideally suited for the placing of control valves. The enzymes involved are subject to a wide variety of effects, many of which depend on the simple, reversible, noncovalent binding of a low-molecular-weight substance to some site in the enzyme molecule. This can be the substrate-binding site, with as consequence an alteration in the accessibility or affinity of this site to the substrate (e.g., competitive inhibition). Or it can be the catalytic site, with consequent modification of its catalytic activity (e.g., noncompetitive inhibition). In a number of particularly important cases, it is a separate site, called allosteric (Greek *allos*, other; *stereos*, solid), whose occupancy induces a change in the conformation of the enzyme molecule that results in a modification of its catalytic activity, or of its affinity for its substrate, or, more subtly, of its affinity for other effectors. Enzymes may also suffer covalent modifications, such as phosphorylation and dephosphorylation, already alluded to (p. 232), which often have drastic effects that completely inactivate or reactivate the affected enzymes. Sometimes, the regulating agents do not act directly on the target enzyme but rather on some other molecule, itself an enzyme or not, that acts as an enzyme modifier, or as a modifier of an enzyme modifier, and so on. The possibilities are endless, and most of them are, in fact, exploited in a large variety of ways. The result is an extraordinarily intricate, multidimensional web of interactions that will challenge for a long time the sharpness of our analytical tools, as well as the power of our imagination—even of our computers.

The main agents of these interactions are normal metabolites or cofactors, internally generated. A special class comprises the actual substrates of the affected enzymes. Such substrate-regulated enzymes have more than one substrate-binding site and are constructed in such a way that occupancy of a site alters for the better (positive cooperativity) or for the worse (negative cooperativity, substrate inhibition) the substrate affinity or catalytic activity of other sites. Such enzymes have anomalous kinetics that differ from the typical Michaelis-Menten type by features that often have profound functional consequences. In a number of cases, the affected enzyme, although not directly concerned with the metabolism of the active agent,

phos unit. What is it that prevents the aerobic cell from recklessly spending ATP, as long as inorganic phosphate is available, in converting glucose into various phosphorylated intermediates? Simple mass-action regulation cannot account for this.

Such examples could be multiplied. They make it abundantly clear that cells have a second set of self-regulative interactions, operating at the enzyme level rather than by simple mass-action effects. The main targets of these interventions are the enzymes that catalyze rate-limiting steps (p. 246) and thereby control metabolic bottle-

Example of feedback inhibition. The
final product of a reaction chain
inhibits the first enzyme of the chain.

nevertheless remains in the family, so to speak, by being part of a chain involved either in the generation or in the breakdown of the agent. Such interactions are often relatively easy to interpret. A typical example is feedback inhibition, a process in which the product of a metabolic chain inhibits the first step of the chain, a straightforward regulating mechanism of the "thermostat" type. Many other interactions, however, cut across metabolic lines and are much more difficult to detect or understand.

In multicellular organisms, a number of particularly potent effects are elicited by exogenous substances that bind to surface receptors. We saw the example of the remarkable cascade of events whereby hormones such as glucagon or epinephrine can, with lightning rapidity, trigger the breakdown of glycogen (p. 232). What was not said at that time is that the cAMP-activated protein kinase that activates the phosphorylase-activating kinase acts also on glycogen synthase, the enzyme that makes glycogen (by transfer of glucosyl groups carried by UDP). But in this case phosphorylation leads to inactivation of the enzyme. One would be hard put to find a simpler and more elegant solution to the futile-cycle problem. Phosphorylation shuts off synthesis and triggers breakdown; dephosphorylation does just the opposite. Not many such examples are known, but there is little doubt that

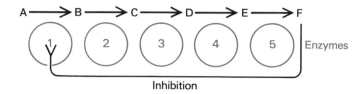

cells use regulating influences to avoid energy waste by futile cycling, except when keeping such cycles going may, like keeping a motor idling, serve some useful purpose deserving of the energy expenditure. This, of course, is not the result of calculated design; it is, like the whole network of interrelationships that give the cytoplasm its dynamic structure and organization, the product of chance trials, tested and screened by natural selection.

In addition to the two levels of regulation that have been considered, there is a third one, which operates by manipulating the number of enzyme molecules. These are generally slow effects in comparison with the others, but they are far reaching in that they may be qualitative as well as quantitative. The dominant mechanisms of this sort are found in the nucleus at the level of gene transcription. We will soon take a look at them and try to assess their significance (see Chapter 16).

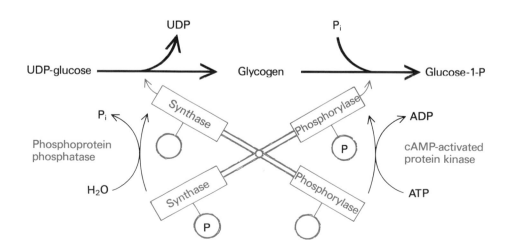

Flip-flop regulation of glycogen breakdown and synthesis by phosphorylation and dephosphorylation, respectively. The phosphorylation of enzymes by cAMP-activated protein kinase activates phosphorylase and inactivates synthase. Dephosphorylation by phosphoprotein phosphatase does the opposite.

Induced Fit, Conformation, Catalysis, and Allostery

On many occasions, in this and in preceding chapters, we have used the lock-key analogy. Born in the imagination of the great German chemist Emil Fischer, superbly adapted to immunology and therapy by his contemporary Paul Ehrlich, this analogy has been immensely useful in explaining such phenomena as antigen-antibody recognition, the interaction between receptors and their ligands, the formation of enzyme-substrate complexes, and the action of hormones and drugs. But like all analogies, especially those that refer to some aspect of the living cell, it is an oversimplification. Proteins are not constructed rigidly like locks; they are very flexible and versatile. To them, admitting a ligand is not the passive acceptance of a key by a lock. It is a real embrace, in which the protein molecule molds itself around its ligand. This does not mean that there is not a high degree of specificity in the interaction. What it means is that the fit is to some extent induced by the ligand, which is equivalent to saying that binding of the ligand causes a conformational change in the protein.

This expression has become something of a catchword, an all-purpose formula glibly invoked for want of a more-precise explanation, a learned-sounding screen behind which we hide our ignorance. But do not let the facile vagueness of the expression make you underestimate its significance. It is a real thing, of immense biological importance, which simply waits to be defined in more-precise detail for each individual protein-ligand complex as our perception of their molecular anatomy sharpens. Receptors, we have seen, depend on this effect to activate their signaling devices. It is the basis of all allosteric regulations. It also underlies all forms of cellular motility. Even in the simple binding of substrates to enzymes, it may play a role in the proper organization of the catalytic site.

Thus, not only at the cell surface, but throughout the cell, ligand-induced conformational changes of proteins represent a universal mechanism of communication, regulation, and catalytic activity, as well as a central means of performing mechanical work. They also provide infinite opportunities for chemical interference with vital processes by means of drugs, poisons, and other pharmacologically active substances that resemble physiological ligands sufficiently to interact in some way with their normal binding sites.

15 | Ribosomes: The Synthesis of Proteins

Our last port of call before leaving the cytosol will be the strings of ribosomes, which, either free or stuck to the surface of ER membranes, spin out the thousands of different polypeptide threads that make up the cell proteins. What this activity means, whether in terms of importance for the cell or of intrinsic complexity, can hardly be overestimated.

The Nature of the Problem

Proteins are by far the most elaborate and versatile substances the biosphere—or, for that matter, the whole of the known universe—has to offer. Throughout our visit, we have seen them performing astounding feats as structural components, enzymes, receptors, gates, carriers, pumps, motor elements, and other functional entities. It is no exaggeration to state that one or more specific proteins lie behind every activity carried out by living cells, or, otherwise put, that life is essentially an expression of proteins. But what then of the processes that manufacture these remarkable substances? Should they not, almost by definition, be even more complex than their products?

The solution to this apparent paradox will become progressively clearer during the last part of our tour. But first let us recognize what the making of a protein molecule represents. On the basis of the chemical information summarized in Chapter 2, we know that all proteins are made of one or more polypeptide chains—that is, strings of linearly assembled amino acids belonging to twenty distinct varieties linked by peptide bonds (—CO—NH—). The innumerable structures shown by the finished products arise from the bending, twisting, coiling, and sometimes joining of such chains and from their fitting with other groups, such as oligosaccharide side chains.

The seemingly insuperable problem of constructing such an immense variety of intricate molecular edifices can be narrowed down to a single dimension, thanks to a central tenet of molecular biology, which states that all the properties of proteins—including their catalytic power, their ability to adopt given configurations and to modify them in response to outside influences, their association tendencies, their susceptibility to glycosylating and other modifying enzymes, and all the other characteristics to which proteins owe their unique biological role—are determined by the primary structure, the amino-acid sequence. In other words, the problem of making a protein is "reduced" (still a formidable job, but at least simpler to formulate) to that of assembling amino acids in the right order. All the rest will follow, provided the necessary ingredients and an appropriate environment are supplied by the cell. For some proteins, this statement needs to be qualified by adding that assembly itself must take place under special conditions. For instance, it is known that certain polypeptide chains undergo changes, such as trimming or glycosylation, while they are being put together. Their final shape is influenced by these changes.

A Mathematical Interlude

Imagine yourself an amino-acid setter. You are sitting in a ribosome and assembling amino-acid letters into polypeptide words. In front of you are 20 boxes, each containing a different kind of amino acid, or a keyboard with 20 keys, or whatever other selection device you may wish. Suppose you are assembling polypeptides containing 100 residues. Such long words do not exist in our language, but, as proteins go, this is a relatively small size, corresponding to a molecular mass of approximately 12,000 daltons. The question is, if you dip into the boxes or hit the keys at random, what are the odds that you will turn up a given sequence?

The answer is a figure not quite as small as the odds of a monkey typing the whole of *Hamlet* by chance, yet small enough to be entirely beyond the boundaries of our imagination. The computation is simple. Your chance of getting the first amino acid right is one in twenty. It is the same for the second one, so that the chances of getting the first two right are 1 in 20^2, or 1 in 400. Repeat this reasoning for every amino acid, and you find that the chances of getting all 100 right are 1 in 20^{100}, or 10^{-130}. Don't bother to visualize this figure or to translate it into familiar terms. Just forget about making polypeptides by chance. Even if the whole population of the earth should be helping you out, working day and night at the incredible speed of 1 million polypeptides per second, never making the same polypeptide twice, it would still take them close to 10^{107} years, or more than 5,000,000,000,000, 000, 000, 000, 000, 000, 000, 000, 000, 000, 000, 000, 000, 000, 000, 000, 000, 000, 000, 000, 000, 000, 000, 000, 000, 000, 000, 000 times the estimated age of the universe, to rattle out all the possible combinations.

Enough; the point is made. Ribosomes cannot work at random; they must be instructed. Which indeed they are, as we shall see, by means of messages from the nucleus. But this does not settle the question entirely. What of prebiotic times, when no instructions were available? According to fossil records, it took less than 1 billion years for recognizable cells, in which hundreds of specific proteins may have been operative, to appear on the surface of our planet. Whatever the mechanisms involved in the assembly of these proteins, it is easily proved that they arose as a result of an infinitesimally small number of trials, compared with the total number of possibilities.

To demonstrate this point, let us make absurdly generous assumptions. The whole mass of the earth (6×10^{27} grams) is involved; it consists entirely of polypeptide chains of 12,000 molecular weight, each represented by a single molecular species, giving $(6 \times 10^{27} \times 6.023 \times 10^{23})/12{,}000 = 3 \times 10^{47}$ peptide molecules. Every billionth of a second the mixture is rearranged to make a new set of polypeptides, never repeating a sequence that has already turned up; this gives us $10^9 \times 60 \times 60 \times 24 \times 365 \times 10^9 = 3.1536 \times 10^{25}$ different sets made in 1 billion years, or $3.1536 \times 10^{25} \times 3 \times 10^{47} =$ about 10^{73} different molecules out of a possible 10^{130}. Even using the entire estimated mass of the universe under these incredible conditions would yield only about one-millionth of the possible sequences in 1 billion years.

Yet the facts are there: the right polypeptides needed to make a primitive cell were actually assembled some 4 billion years ago, with a total number of trials that cannot but have been vanishingly small with regard to the number of possibilities. Was it, as some biologists believe, a fantastic stroke of luck, a unique event, never to be repeated anywhere in the universe? Such a belief cannot be disproved, but it defies the laws of probability. The odds of getting one polypeptide right are small enough, 10^{-130}, to make the event close to impossible. How, then, should we qualify the probability of getting 100 different polypeptides right at the same time, or $10^{-13,000}$? To beat such odds, no less than a miracle is required.

What then? Do we have to assume that the prebiotic assembly of polypeptides was "instructed," guided by an unseen hand, so as to produce sequences possessing the catalytic properties required to put life on course? Not necessarily. There is an alternative possibility: namely, that catalytically active polypeptide sequences are the commonest thing on earth; that, in fact, almost any polypeptide has some enzymic activity. Once we make this assumption, all we need are conditions such that amino acids will arise and assemble spontaneously. If this happens, the polypeptides formed are bound to include primitive enzymes of various kinds, even though only an infinitesimally small proportion of the possible molecules has been generated. This is exactly what has been found in experiments designed to reproduce in the laboratory conditions believed to have prevailed on earth in prebiotic times. Authentic amino acids have indeed been obtained from such commonplace ingredients as methane, ammonia, hydrogen, and water. Primitive "proteinoids" have been produced from amino acids by heat or other simple physical means, and these crude artificial polymers, clearly created without the benefit of any sort of instruc-

tion, have been found to display crude catalytic properties similar to those of enzymes.

We will revert to this fascinating topic in Chapter 18. First, we must look at the protein factories of today. Whatever happened 4 billion years ago may remain forever a matter for conjecture. What we do know, however, is that it resulted in the development of a biosynthetic system capable of putting amino acids together in specified sequences that faithfully reproduce those that exist in the cell.

The Cipher of Life

Should we be asked to guess how a ribosome can possibly "know" which of the twenty available amino acids to add at each step while assembling a polypeptide chain, we would no doubt think in terms of copying. We would visualize ribosomes using the cell's proteins as templates and painstakingly reproducing them, amino acid by amino acid, with the care and accuracy of a medieval copyist. Asked to provide a chemical mechanism for such a process, we might invoke the attraction of like for like, the force behind crystallization, as did the American biochemist Felix Haurowitz not much more than 30 years ago. We would be wrong. The instructions come in code, delivered by a completely different type of molecule that belongs to the class of ribonucleic acids (RNA) and is appropriately named messenger RNA (mRNA). What happens is not copying, but translation.

Ribonucleic acids share certain features of polypeptides. They are long, garlandlike macromolecules, all with the same iterative backbone made of n identical molecular units and rendered uniquely informational by specific side groups, or "letters," attached to the backbone. But here the analogy ends. The backbone is made of a coarser fiber, quite unrelated to that of polypeptides. The side groups also differ from those of polypeptides. And, especially, there are only four of them. The RNA alphabet is much poorer than the protein alphabet: four letters (AGUC), against twenty.

Chemically, RNAs are defined as polynucleotides, made by the linear association of mononucleotide units. We have already encountered these building blocks, which are the 5'-monophosphates of nucleosides. In RNA, these units are linked by phosphodiester bonds that attach the terminal phosphate of each nucleotide to the 3'-hydroxyl group of its neighbor:

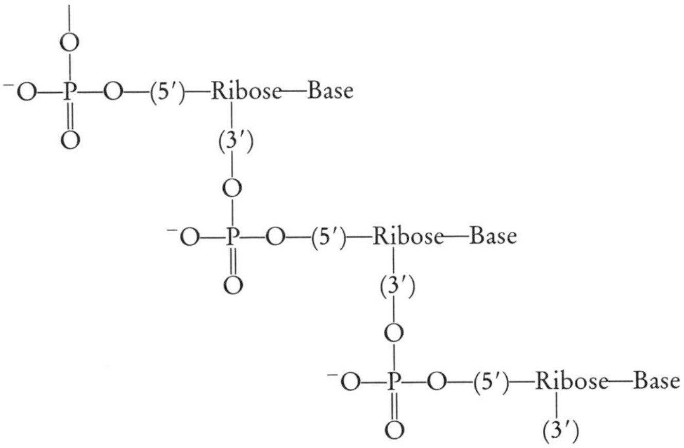

The backbone consists of 5'-phosphoryl-ribosyl units, n times repeated. At one end, called the 5' end, the 5'-phosphoryl ester group is not engaged in a phosphodiester linkage. At the 3' end, the 3'-hydroxyl group of the last ribose is free. The four bases are the two purines, adenine and guanine, and the two pyrimidines, cytosine and uracil, already mentioned.

It is useful to point out at this stage that DNA has essentially the same structure, except that it has a different pentose (5-carbon sugar) in its backbone—deoxyribose—and uses thymine instead of uracil as one of its pyrimidine bases; its alphabet is AGTC. But these are minor differences: deoxyribose is simply ribose without an oxygen atom in position 2; thymine is uracil with an additional methyl group in position 5 (see Appendix 1). Information transfer is not affected by this difference. The two alphabets are essentially the same: T and U are interchangeable.

Nucleic acids, both RNAs and DNAs, have one additional feature of absolutely cardinal importance that is not

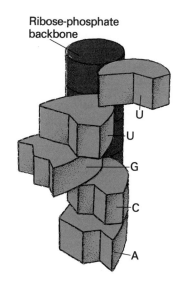

Ribose-phosphate backbone

immediately evident unless you look at them the right way: their constitutive bases are complementary two by two, the purine adenine with the pyrimidine uracil (RNA) or thymine (DNA), and the purine guanine with the pyrimidine cytosine (both RNA and DNA). The chemically minded among you can see from the structures of the two partners (facing page) how base-pairing takes place: A binds to U or T by means of two hydrogen bonds (Chapter 2), and G to C by means of three, making the GC pair the stronger of the two.

Those who are frightened by chemical formulas may visualize the paired bases as two flat pieces, joined by perfectly fitting edges to form a rigid planar structure of roughly elliptical shape, with axes of 1.1 and 0.6 nm and a thickness of 0.34 nm. A cardinal consequence of base-

Chemical mechanism of base-pairing. The top diagram shows how the purine base adenine (A) joins with the pyrimidine base uracil (U, R=H) or thymine (T, R=CH₃) by means of two hydrogen bonds. The lower diagram shows similar joining between the purine guanine (G) and the pyrimidine cytosine (C) by means of three hydrogen bonds. Each base pair forms a flat plate, roughly elliptical in shape, with axes of 1.1 and 0.6 nm, and a thickness of 0.34 nm. The pentose-phosphate groups attached to the bases can rotate freely around the glycosidic N-C bond. But, when base-pairing occurs between two nucleic-acid chains (RNA-RNA, DNA-DNA, or RNA-DNA), these groups are forced to assume the positions indicated in the diagram, with 3′ ends pointing in opposite directions. Grooves refer to the double-helical structure of complementary chains (see the drawing on p. 256).

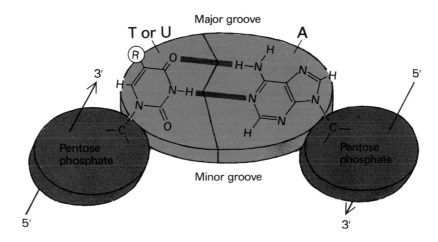

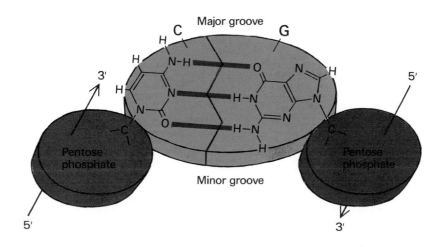

Regular double-helical structure generated by perfectly complementary nucleic-acid chains. Note the minor and major grooves and the double row of negative charges on the outside of the backbones. Several different structures are possible. The diagram illustrates the B form, with a pitch of 3.4 nm (or ten base pairs) per complete turn, first proposed for DNA by Watson and Crick in 1953 (see Chapter 16) and characteristic in first approximation of most of the DNA in nature.

This entrance to the Vatican museum has a pair of staircases in the form of a double helix, one staircase for going up, the other for going down.

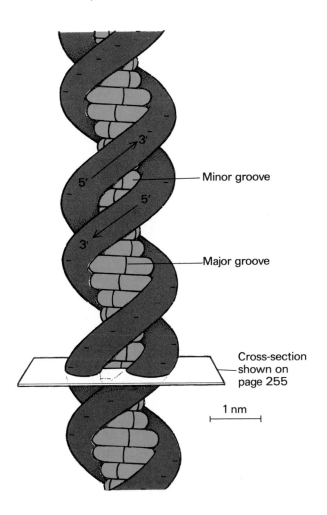

Minor groove

Major groove

Cross-section shown on page 255

1 nm

pairing is that it may serve to link two nucleic-acid strands, provided the two backbones run antiparallel to each other (one in the 5′——→3′ and the other in the 3′——→5′ direction). The longer the complementary sequences, the stronger the link. Owing to the characteristic molecular anatomy of polynucleotide chains, such arrangements tend to adopt a regular helicoidal shape. The paired bases pile up in parallel planes, but with their long axes forming an angle of 36°, somewhat like the steps of a spiral staircase or, rather, a double spiral staircase, one for going up, the other for going down, such as exists at one of the entrances to the Vatican museum in Rome. The two antiparallel backbones run along the outer edges of the base pairs, like the banisters of the two staircases.

When it extends over ten or more complementary base pairs, this structure appears as a twisted cord about 2 nm thick but narrowed by two helical grooves, a deep one and a shallow one. Its turn is right handed and its length is 0.34 nm per base pair, making its pitch 3.4 nm per com-

Diagrammatic representation of the formation of a loop in RNA. The loop is closed by the pairing of two short complementary stretches in anti-parallel orientation.

plete turn ($10 \times 36°$, the angle between adjoining base pairs). A double-helical row of negative charges (20 per turn), provided by the phosphoryl groups of the two backbones, lines the outside of the cord.

The most perfect double helices are found when the two strands are complementary from one end to another, as is the case with most DNA molecules and, exceptionally, with some double-stranded viral RNAs. Most RNA molecules are single stranded but include, for reasons that are obvious in view of the limited number of different combinations allowed by their alphabet, many short base sequences that are complementary in antiparallel fashion to sequences occurring further along the same strand. Pairing of these sequences causes the formation of loops closed by double-helical segments of variable length; for example:

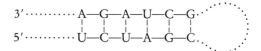

Thus, like proteins, most RNAs form characteristic secondary structures stabilized by interactions between side chains. When represented schematically, these structures resemble graceful mobiles, which have been described poetically as cloverleaves, flowers, and the like. In reality, most of them must appear as hideously twisted knots. As in proteins, such arrangements may be denatured by heat or other agents to generate a random tangle of filaments.

But the function of base-pairing is not simply a structural one. Its main role is communication. Amazingly, these two elementary relationships

$$\boxed{A=U} \quad \text{or} \quad \boxed{A=T} \quad \text{and} \quad \boxed{G\equiv C}$$

govern, through the two relatively fragile structures they embody, the whole of information transfer throughout the biosphere. They are truly the cipher of life. This will become evident as we look at ribosomal translation, and even more so when we learn about transcription and replication in the nucleus.

The Code

Messenger RNAs specify the sequence of amino acids in polypeptides co-linearly, with the $5' \longrightarrow 3'$ direction corresponding to the N-terminal $\longrightarrow$ C-terminal direction in the polypeptide. In other words, the signs representing the amino acids follow each other, unbroken and unscrambled, from the $5'$ to the $3'$ end of the mRNA in the order in which the amino acids themselves follow each other in the polypeptide from the N-terminal to the C-terminal end. The message does not, however, start at the $5'$ end and stop at the $3'$ end of the mRNA; it is preceded and followed by mute sequences that are not translated. As to the message itself, since the RNA alphabet contains only four letters it is obvious that several letters must be used to code for each amino acid. How many is easily computed: with n letters in the alphabet you can construct n^m different words of m letters.

Number of letters in alphabet	Number of words in vocabulary			
	One-letter	Two-letter	Three-letter	Four-letter
2	2	4	8	16
4	4	16	64	256
20	20	400	8,000	160,000
26	26	676	17,576	456,976

By looking at the table above, you can readily see that with a four-letter alphabet (RNA language) you need words of at least three letters to represent all of the twenty amino acids (protein language), at least if you want all your words to be of equal length. You could just get by with less by using words of different lengths, as is done with the two-letter alphabet of the Morse code; but this was not Nature's way. You will also notice that with three-letter words you suffer from an *embarras de richesse*: sixty-four combinations to convey twenty meanings. You could, as in our own language, use only part of the possible combinations and reject the others as nonsense. But, again, this was not Nature's way. All sixty-four possible combinations of three bases are used in the genetic code.

The breaking of this code has been one of the great triumphs of modern biology. It started in 1961, when a young American investigator discovered that an artificial polyribonucleotide containing only uracil (poly-U) induced the synthesis of a polypeptide made only of phenylalanine (poly-Phe). This yielded the first line of the genetic dictionary: UUU = Phe. Five years later the whole dictionary had been elucidated. Now we use it not only to read, but even to write, genetic messages and, with the help of genetic engineering techniques, to instruct cells to translate the message for us in as many copies as we wish (see Chapter 18). The code used by the great majority of living beings is reproduced in the table on the facing page.

As you can see, the genetic dictionary is a highly ordered system. All amino acids, with the exception of Met

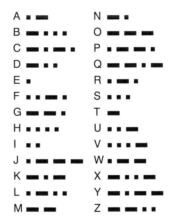

The international (continental) Morse code represents the 26 letters of the alphabet by most of the 30 combinations of two distinct signs (short-long) that can be made with codons of between one and four signs.

and Trp, are represented by more than one base triplet, or codon. Some (Leu, Ser, Arg) have as many as six synonyms. In all cases, replacing U by C at the end of the codon does not alter its significance. Often, but not always, A and G are also interchangeable in third position. The third base is entirely irrelevant for eight out of the twenty amino acids. Altogether, sixty-one triplets code for amino acids. The remaining three (UAA, UAG, and UGA) are stop signals. As we shall see, AUG, and sometimes GUG, have a special additional role as initiators.

Why this particular code, and not another? This question has already caused much ink to flow. But before addressing it, we must bear in mind that there are a few dissenters. Among them, our own mitochondria. For

Second letter

		U	C	A	G	
First letter	U	UUU } Phe UUC UUA } Leu UUG	UCU } UCC } Ser UCA } UCG	UAU } Tyr UAC UAA OCHRE† UAG AMBER†	UGU } Cys UGC UGA OPAL† UGG Trp	U C A G
	C	CUU } CUC } Leu CUA } CUG	CCU } CCC } Pro CCA } CCG	CAU } His CAC CAA } Gln CAG	CGU } CGC } Arg CGA } CGG	U C A G
	A	AUU } Ile AUC AUA AUG* Met	ACU } ACC } Thr ACA } ACG	AAU } Asn AAC AAA } Lys AAG	AGU } Ser AGC AGA } Arg AGG	U C A G
	G	GUU } GUC } Val GUA } GUG*	GCU } GCC } Ala GCA } GCG	GAU } Asp GAC GAA } Glu GAG	GGU } GGC } Gly GGA } GGG	U C A G

Third letter

*These codons have a second function as initiators of translation.
†These codons are terminators, or stop signs; UGA is read as Trp by mitochondria. For other peculiarities of the mitochondrial code, see text.

them, UGA does not mean stop, but rather Trp, like its close neighbor UGG. This peculiarity adds to all the others that support a separate origin for mitochondria, perhaps from an ancestral bacterial endosymbiont, especially since other mitochondria (e.g., of yeast and of the fungus *Neurospora*) also read UGA as Trp. Something as simple as a separate universal mitochondrial code does not exist, however. Whereas our mitochondria and those of *Neurospora* read CUA as Leu, in the orthodox way, yeast mitochondria deviate by making it Thr. On the other hand, ours have the distinction of understanding AUA as an order for Met instead of Ile.

On the whole, however, the biosphere is no Babel. It is, with the rare exceptions just discussed, linguistically homogeneous. Viruses, bacteria, fungi, plants, animals, and humans all use the same code. Does this mean that our code is the only possible code, imposed by a set of unique chemical relationships between the amino acids and their respective codons? Or, on the contrary, is the assignment purely random, one that just occurred by chance in the primitive cell from which all living organisms are believed to have originated? Neither one nor the other, apparently. Amino acids, as we shall see, have nothing to do with the reading of their codons and are not likely to have had much say in their choice, or vice versa. On the other hand, the code has an obvious structure; it is not random. It is quite clear, for instance, that chance could not possibly have grouped synonyms the way they

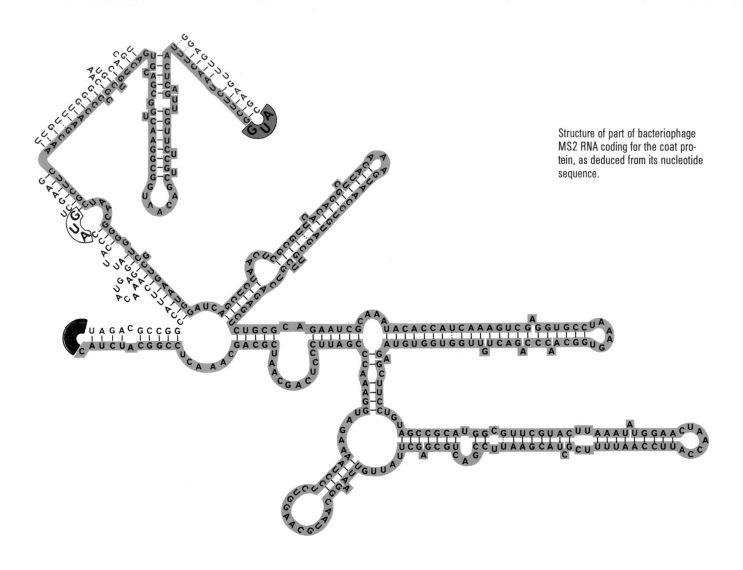

Structure of part of bacteriophage MS2 RNA coding for the coat protein, as deduced from its nucleotide sequence.

are. Their spelling similarities have suggested that in early prebiotic times there may have been no more than sixteen amino acids, each coded for by a doublet followed by an essentially meaningless third base. Some of these codons were then "borrowed" to code for the additional amino acids that appeared and to serve as stop signs. It has also been pointed out that the code is such as to minimize translational errors due to misreading, as well as deleterious effects of point mutations involving the replacement of one base by another, and may therefore be the product of natural selection. Indeed, you can easily verify that many base replacements either do not change the meaning of codons or, if they do, substitute an amino acid that resembles its predecessor in certain key properties

(hydrophobicity, electric charge), so that the conformational and functional properties of the altered protein are not drastically affected. These evolutionary aspects will be considered further in Chapter 18.

When the genetic code was first deciphered, only a few proteins had been sequenced and techniques for sequencing nucleic acids were still in their infancy. The early cryptologists achieved success by indirect means through ingenious experimentation. Today, sequencing methods have been improved to such an extent that even very long messages can be "read" directly, and compared with their translation products.

One of the first natural messages to be completely decoded in this way is the RNA of a small bacterial virus, or

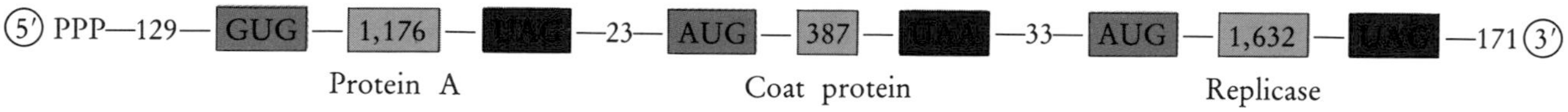

(5') PPP—129— GUG — 1,176 — UAG —23— AUG — 387 — UAA —33— AUG — 1,632 — UAG —171 (3')

            Protein A   Coat protein   Replicase

bacteriophage, called MS2. Made of 3,569 nucleotides, this RNA doubles up in more than sixty loops, stabilized by complementary base sequences, to make up a highly intricate "flower" structure, which, in three-dimensional reality, must appear as the Gordian nightmare par excellence. Genetically, this RNA codes for three proteins (sometimes for four, but we won't consider this complication): protein A, coat protein, and the enzyme replicase. The first two, but not the third, have been sequenced. The insertion of the three corresponding messages in the RNA is given schematically above, with only initiation and termination codons shown explicitly and the other sequences indicated simply by the number of nucleotides involved.

Looking at the messages first, we see that two start with AUG, one with GUG; two are terminated by UAG, one by UAA; all according to the book. In protein A and the coat protein, which have been sequenced, message and translation product were found to conform perfectly to the code. The sequence of the replicase, which was not known by direct analysis, could be deduced from that of the corresponding RNA message. The messages themselves are flanked by mute (untranslated) sequences of variable length. The 5' end is "capped" by a pyrophosphate group attached to the terminal phosphate, reflecting

that the primary acceptor in RNA assembly is an NTP (see Chapter 16). These details are not useless, but serve to lure the infected microbe into accepting the viral message and to position it correctly in its translation machinery, as though it were one of its own. This delusion proves fatal to the bacterium and highly beneficial for the spreading of the virus, since the A and coat proteins are the two protein components of the virus and the replicase is an enzyme needed for the replication of RNA. Thus, by getting its message accepted for translation, the virus actually subverts its unfortunate host into devoting its resources to making new virus particles. The multiplication of viruses always relies on some subversion of this kind, although a number of different mechanisms may be involved. More will be said about viruses in Chapter 18.

Eukaryotic mRNAs also include mute sequences. Especially important is the "leader" sequence that precedes the message at the 5' end and ensures a proper initiation of translation (see p. 270). With rare exceptions (e.g., histone mRNAs), the 5' end of eukaryotic mRNAs bears several added methyl groups plus a characteristic 7-methyl-guanosine triphosphate cap, and the 3' end is terminated by a long poly-A tail of between 100 and 150 A units. Thus, a typical eukaryotic mRNA has the following structure:

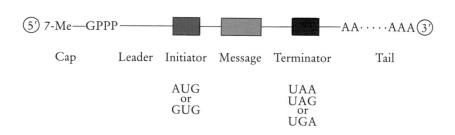

(5') 7-Me—GPPP————————————————AA·····AAA (3')

Cap   Leader   Initiator   Message   Terminator   Tail

                           AUG   UAA
                            or   UAG
                           GUG   or
                                            UGA

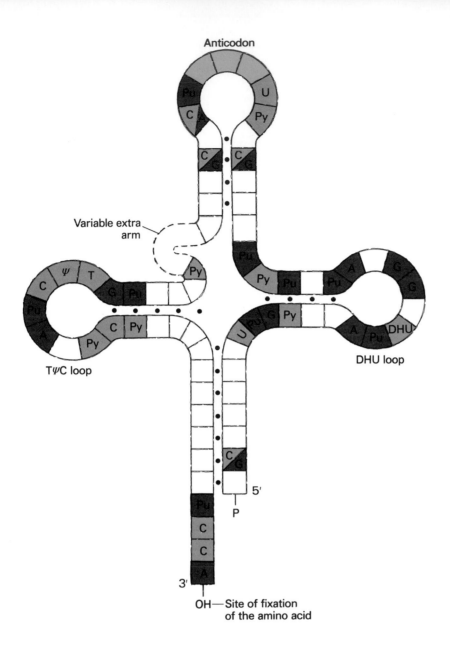

Anticodon

Variable extra arm

TΨC loop

DHU loop

5′

P

OH—Site of fixation of the amino acid

3′

Structural characteristics common to all or most tRNAs. DHU (dihydrouridine) and Ψ (pseudouridine) are modified nucleosides. Pu stands for purine; Py for pyrimidine. C/G means C or G; C/A means C or A. The number of nucleotides in each loop may vary somewhat from one type of tRNA to another.

Reading the Message

Amino acids, we have just seen, do not read their codons. A number of attempts have been made to discover chemical affinities between the side chains of the various amino acids and the base doublets or triplets that code for them. But to no avail. The protein and RNA languages seem unrelated. How, then, is the message read? Here is where our cipher comes in. Each amino

acid is attached to a special kind of RNA called transfer RNA (tRNA), which serves as reading device through base-pairing. These tRNAs are remarkable molecules, small as RNAs go (about 25,000 molecular weight) and made up of some seventy-five to ninety-three mononucleotide units, but with many of their bases modified by methylation, reduction, deamination, or transposition of the nucleosidic linkage (see Appendix 1). They all have a -CCA 3′ terminal sequence, together with certain strate-

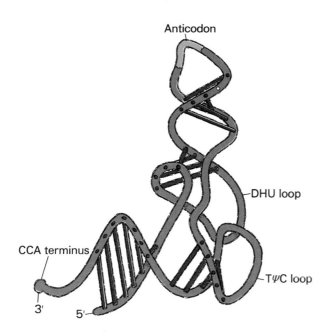

Anticodon

DHU loop

CCA terminus

TψC loop

3'   5'

Skeletal representation of the true conformation of yeast tRNA^{Phe} (transporting the amino acid phenylalanine). Compare with the schematic diagram on the facing page. Note the double-helical segments made by base-pairing.

gically placed nucleotides, and they all possess four pairs of complementary base sequences situated in such a way as to give the molecule a "cloverleaf" shape. Actually, the real structure of tRNAs is rather different from that shown in the diagram. Because of the double-helical torsion of the paired sequences that close the loops, it is screwed up into a tight knot. Even so, two features are clearly distinguishable: sticking out at the 3' end is the typical CCA terminus with its free 3'-hydroxyl group, which serves as acceptor for the amino acid; and, diametrically opposed, occupying the exposed part of the middle loop, is the reading device, or anticodon, a base sequence complementary to the codon.

This complementarity is strict for the first two bases of the codon but may be "wobbly" for the third. For example, the anticodon of brewer's yeast tRNA^{Asp} (the tRNA carrier of aspartic acid) is CUG, read from 3' to 5'. This gives a perfect match with one of the codons of aspartic acid and an imperfect match, through the first two bases of the codon, with the other:

$$\text{Codon } (5' \longrightarrow 3') \quad \text{—GAC—} \quad \text{—GAU—}$$
$$\text{Anticodon } (3' \longleftarrow 5') \quad \text{—CUG—} \quad \text{—CUG—}$$

The second example illustrates a case of wobble: recognition is mediated by the first two bases; the third one is loose. Thanks to the wobble, it is possible for some anti-codons to recognize two or three, but usually no more than three, different codons specifying the same amino acid, allowing cells to get by with between thirty-five and forty different tRNAs to recognize the sixty-one amino-acid codons.

Recognition, you will notice, is done entirely in RNA language, a fact that has been elegantly verified by artificially attaching the "wrong" amino acid to a tRNA: the substitute is not recognized as such and is handled exactly as if it were the right amino acid. As we shall presently, in the assembly of a polypeptide the amino acids are lined up one by one along the mRNA by means of their tRNAs; their correct insertion is determined exclusively by the accuracy and specificity of codon–anticodon interactions, and hence by the implied correctness of the amino-acid–tRNA association formed earlier.

It follows that the actual step of translation from RNA into protein language occurs when amino acids and tRNAs are matched and joined. The translators are the enzymes. They are the only bilingual elements in the cell: they can recognize both an amino acid and some part of its corresponding tRNAs, which, however, does not seem to be the anticodon itself. Interestingly, each enzyme recognizes all the tRNAs of a given amino acid. Therefore, it has imprinted in its structure one line of the genetic dictionary, with all synonyms included. There are twenty such enzymes, one for each amino acid. Together they make up the complete dictionary. But they do so in a cryptic form that relies on the tRNAs for decoding into anticodon language.

These scholarly enzymes double as energizers. They belong to the group of ligases and, in addition to their crucial role as translators, serve to provide the amino acids with the energy needed for closing the peptide bond. We encountered them before when we looked at the activation of biosynthetic building blocks. As mentioned earlier, peptide synthesis is a three-step, sequential, group-transfer process dependent on a β_p attack of ATP and on the participation of tRNA as carrier of the aminoacyl group. This is all the information we need to reconstruct the event, with the help of the general notions explained in Chapter 8.

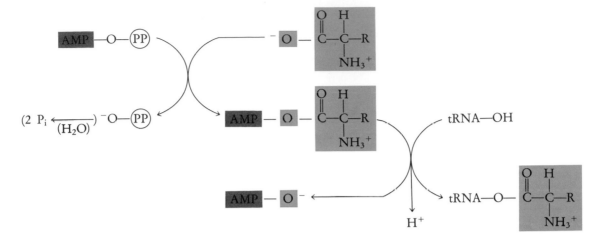

Steps 1 and 2 (activation), catalyzed by an AMP-forming ligase, occur by successive AMP-yl and aminoacyl transfer, with aminoacyl-AMP serving as double-headed Janus and oxygen transmitter from the amino acid to AMP. As is the rule in such cases, the PP$_i$ formed is hydrolyzed to P$_i$, as shown above.

The third step, which takes place on ribosomes, transfers the activated aminoacyl group from its tRNA carrier to its final acceptor, an amino group belonging to an amino acid. But there is a twist to this reaction. The aminoacyl group is not, as might be expected, transferred immediately to its final acceptor. It first accepts a growing polypeptide chain on its amino group and only subsequently is transferred from its tRNA carrier with use of the aminoacyl-tRNA bond energy for the synthesis of a peptide bond. A protein grows by its C-terminal head, not by its N-terminal tail (Chapter 8), as shown schematically below.

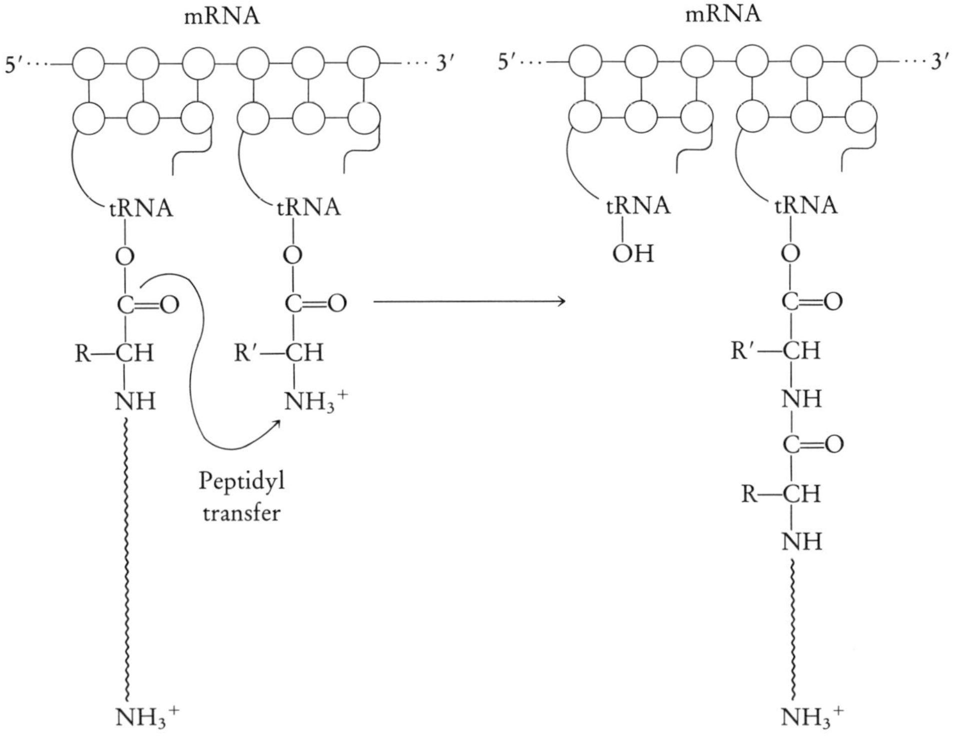

As indicated, the growing tRNA-linked chain remains attached by its tRNA to the mRNA, which is read, codon by codon, in the $5' \longrightarrow 3'$ direction, while the polypeptide is lengthened in the N-terminal$\longrightarrow$C-terminal direction.

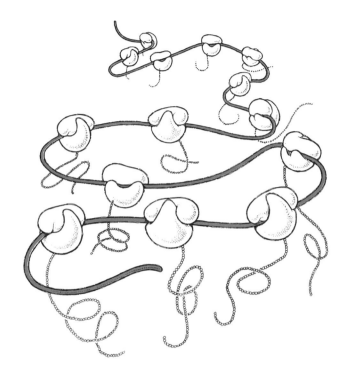

The Perfect Robot

So much for our preliminary briefing. We can now resume our visit and start looking for some real ribosomes so that we can watch them in the act. They are not hard to find. The average mammalian cell contains some ten million ribosomes, and they are all, except for the small number of special ribosomes present in the mitochondria, directly accessible from the cytosol. A good many of them are attached to the surface of ER membranes. But others are in the cytosol, and these are the ones we will look for since they are easier to observe.

Seen from some distance, ribosomes appear as small, compact particles, slightly oblong in shape, with overall dimensions of between 15 and 25 nm—about the size of a pigeon egg at our magnification. They are not randomly distributed but are characteristically clustered by groups of ten or more. As we approach them, the reason for this clustering becomes obvious: the particles are linked, like beads on a string, by a long, slender thread. A typical 7-methyl-GTP cap at the 5' end and a long tail of poly-A at the 3' end identify this thread as an mRNA.

The ribosomes themselves are made of two pieces of unequal size: the large and small subunits. The large subunit looks like a stubby pear with much of its top part bitten off. The small subunit, about half the weight of the large one, is an asymmetric, bean-shaped piece lying transversely across the large subunit. The two subunits fit snugly together, except where a cleft in the small subunit leaves a channel or tunnel open between them.

The main chemical constituent of ribosomes is RNA (hence the name *ribo*somes), which makes up about 55 per cent of their weight. The remainder consists of protein. The ribosomal RNAs (rRNA) are represented largely by two unbroken strands, about 2,000 and 5,000 nucleotides long, that occupy the small and large subunits, respectively. The large subunit contains in addition two RNAs of smaller size. In biochemical jargon, the two long rRNAs are usually referred to as 28S and 18S (23S and 16S in bacterial ribosomes, which are smaller), and the short ones as 5.8S and 5S. These measurements are derived from the rate at which the molecules sediment in a centrifugal field and are given in Svedberg units, named after the Swedish scientist who developed the technique of analytical centrifugation for the determination of the molecular weight of macromolecules.

Some seventy-five to eighty different proteins (fifty-three in prokaryotes) have been identified in ribosomes, thirty in the small subunit and from forty-five to fifty in the large subunit. Many of them have been purified and sequenced, especially in the simpler prokaryotic ribosomes, and much is already known concerning the manner in which they combine with each other and with the rRNAs to form the two ribosomal subunits. Soon the complete molecular anatomy of a ribosome of the intestinal bacterium *Escherichia coli* will be known, to provide us with the most elaborate structure of a biological organelle as yet unraveled, a unique construction of more than half a million atoms. When mixed together under appropriate conditions, the sixty-odd pieces that make up a bacterial ribosome recombine spontaneously into a functional particle by self-assembly. This, no doubt, is how ribosomes arise inside the cell: their complete blueprint is contained in the primary structure of their constituents.

At the center of the ribosomal edifice lies the enzyme peptidyl transferase, whose duty it is to close the peptide bond at each successive step of polypeptide assembly. The whole intricate scaffolding around it is there to ensure the accurate positioning (and repositioning after each bond is sealed) of the mRNA and of the two tRNA-linked reactants, all the way from initiation to completion of the chain. It is assisted in this job by a number of soluble helper proteins—initiation factors, elongation factors, release factors, at least three of each—and by a liberal supply of energy provided by the hydrolysis of GTP to GDP and P_i. Let us watch this remarkable machinery as it does what it does most of the time: elongating a polypeptide chain.

Running across the ribosome, bound to the small subunit near its junction with the large subunit, lies the mRNA, the tape containing the instructions for the assembly process. Note that the specific nature of the product is entirely dictated by this tape. The rest of the machinery is strictly standard. It will turn out any kind of polypeptide, just as your cassette player will play any tune for you if properly instructed. As we face it, the 5′ end of the mRNA lies at our left, the 3′ end at our right. Therefore, the part of the message that has already been translated is to the left; what remains to be translated is to the right.

The growing polypeptide chain is readily identified by its free-floating N-terminal end. As we move up from this

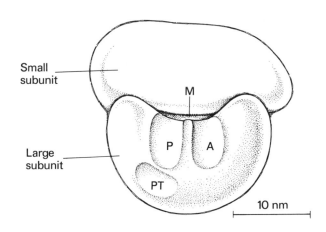

Scematic funtional map of a ribosome showing the site for binding of mRNA (M), the peptidly site (P), the aminoacyl site (A), and peptidyl transferase (PT).

end along the silky filament, we recognize the amino-acid residues in the order of their successive attachments until we reach the last one, which provisionally makes up the C-terminal head of the growing chain. It is linked by its carboxyl group to the 3′ end of its tRNA, and this tRNA is itself joined by its anticodon to the corresponding codon in the mRNA. The whole bulky structure is tightly secured within a complex housing that spans the two ribosome subunits and is called the peptidyl, or P, site. This site is adjacent to the central peptidyl transferase and is organized in such a way as to put the terminal aminoacyl group of the growing chain in close contact with the active center of this enzyme. Immediately to the right of the P site lies another irregularly molded cavity of very similar shape, connected with the next codon to the right on the mRNA. It is called the aminoacyl, or A, site, and is empty at present. The stage is set for the ribosome to perform its favorite act.

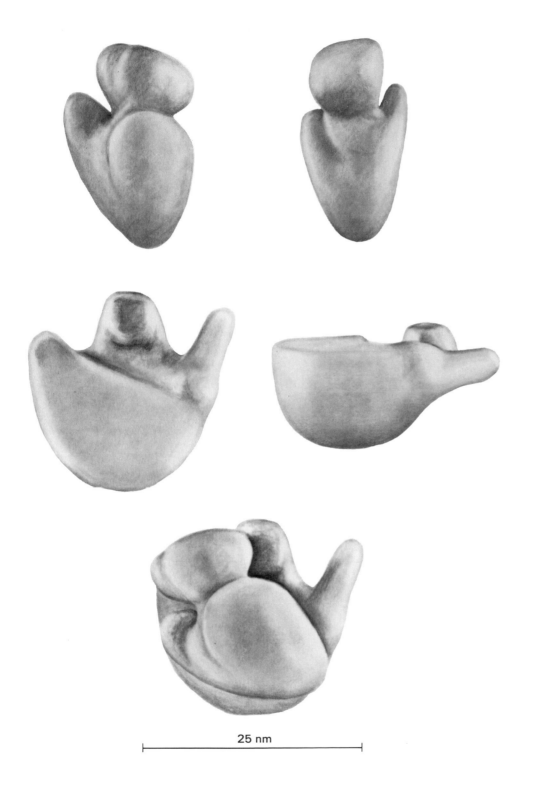

Model of *E. coli* ribosome. Top: two views of the small subunit. Middle: two views of the large subunit. Bottom: view of the assembled ribosome.

25 nm

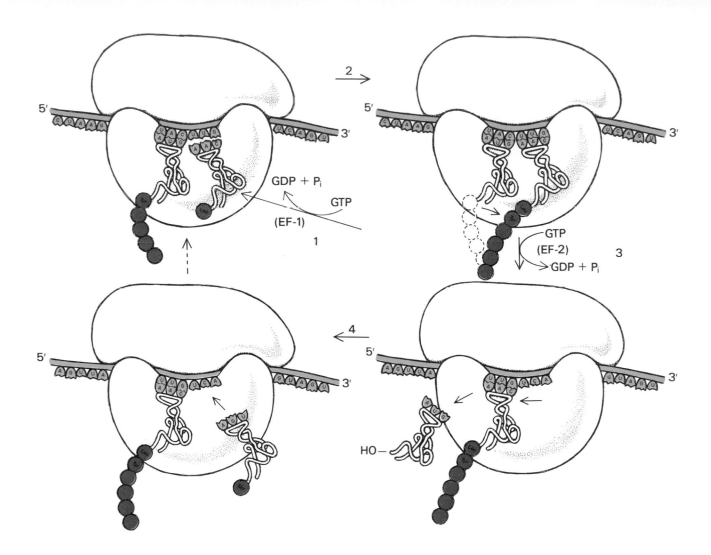

Protein synthesis: chain elongation.

Step 1: With the help of elongation factor EF-1 and the hydrolysis of one GTP molecule, the correct aminoacyl-tRNA is inserted into the A site of the ribosome (which has its P site occupied by the growing peptide chain attached by means of a C-terminal tRNA).

Step 2: The peptidyl transferase transfers the entire growing chain to the amino group of the aminoacyl-tRNA.

Step 3: With the help of elongation factor EF-2 and the hydrolysis of another GTP molecule, the lengthened peptidyl-tRNA is translocated from the A site to the P site, dislodging the free tRNA and carrying along the mRNA, which now offers a new codon to the aminoacyl site.

Step 4: The cycle starts again, with a new aminoacyl-tRNA corresponding to the codon above the A site.

For the act to proceed, the right aminoacyl-tRNA must be inserted into the A site. There may be as many as forty different kinds of aminoacyl-tRNAs swimming around, so quite a number of trials may take place before a correct codon-anticodon fit is obtained. (Remember, this fit is the only way whereby correct identification of the next amino acid to be attached can be accomplished.) Helping in this critical binding-recognition process is an important elongation factor called EF-1, which serves as carrier for the various aminoacyl-tRNAs and, by using its own binding sites on the ribosomal surface, presents the aminoacyl-tRNAs to the A site in the appropriate orientation for fitting to take place. This EF-1 factor also carries a GTP molecule. If you can adjust your eyesight to

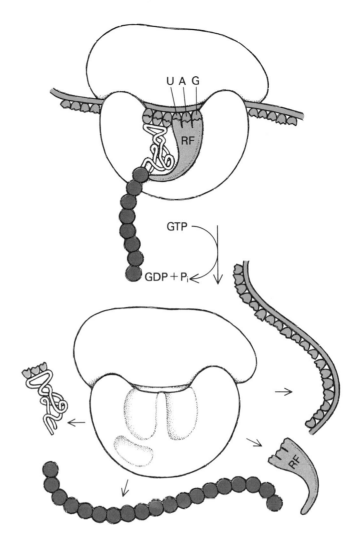

Protein synthesis: chain termination. Upon the appearance of the termination codon UAG (or UAA or UGA) over the A site, releasing factor RF, with the help of the hydrolysis of one GTP molecule, causes the release of the completed polypeptide chain, with the concomitant liberation of the tRNA and detachment of the mRNA from the ribosome.

the whirl of molecular agitation, you may just be able to distinguish the multiple random trials that occur until the right anticodon aligns against the codon bordering the A site. As it snaps into place, something happens to the contact area where EF-1, GTP, and ribosome meet; the GTP is split and EF-1 floats away, leaving the aminoacyl-tRNA correctly inserted into the A site. This implies appropriate codon-anticodon pairing at the mRNA end of the A site and positioning of the free amino group of the aminoacyl radical close to the active site of the peptidyl transferase.

That is all this enzyme has been waiting for in order to do its job, which is transferring the whole growing peptide chain from its tRNA carrier to its NH_2 acceptor. As a result of this transfer, the chain becomes longer by one residue and is now attached to the A site by means of its newly acquired C-terminal amino acid and this molecule's tRNA. This is the signal for a second elongation factor, EF-2, to take over. With the help of the hydrolysis of another GTP molecule, it chases away the free tRNA that occupies the P site and causes the ribosome to translocate the whole mRNA-tRNA-peptidyl complex from the A site to the P site. We are back at stage one; the elongation act can start again.

Ribosomes go through this routine tirelessly, all ten million of them, obeying thousands of different messages and going "clickety-click" several times per second. Usually, a dozen or more ribosomes follow each other along the same mRNA strand, moving from the 5' end to the 3' end at the rate of ten to fifteen nucleotides per second, each trailing along a regularly lengthening polypeptide thread. We had a glimpse of this scene in Chapter 6 when we watched it through the dimly transparent membranes of the ER. There is something nightmarish about it, straight out of Charlie Chaplin's *Modern Times*, as we see it now in close detail, going on all over the cytosol and its membranous partitions endlessly, mindlessly, with metronomic regularity. But there is also great beauty in the efficiency of the machine, finely honed and perfected by natural selection, and in its excellent record of accuracy: probably no more than one mistake in ten thousand; the equivalent of one typo per five to ten pages of your average paperback. What typesetter could claim anything comparable? Mistakes in protein synthesis, incidentally, are generally of little consequence; they just cause a few faulty models of a given molecule to be produced and, since the errors vary more or less at random, no more harm than slight wastage of energy can result.

Two events break the drudgery of a ribosome's life: initiation and termination. The latter happens when one of the three stop codons—UAA, UAG, or UGA—slips into place over the A site at the last translocation, to be recognized by no tRNA (unless some mutation has modified the recognition pattern). These codons are recognized by releasing factors, which, again at the expense of one GTP, cause the successive release from the ribosome

Protein synthesis: chain initiation. Initiation factor IF, with the help of the hydrolysis of one GTP molecule, causes (1) the dissociation of a ribosome into two subunits; (2) the formation of an "initiation complex" between the small ribosomal subunit, the 5′ end of an mRNA (positioned so as to display an initiation codon AUG (or GUG) above what will become the P site), and a special tRNA bearing a methionyl group (formylated in bacteria); (3) association of this complex with a large subunit. Elongation can now start (see the illustration on p. 268).

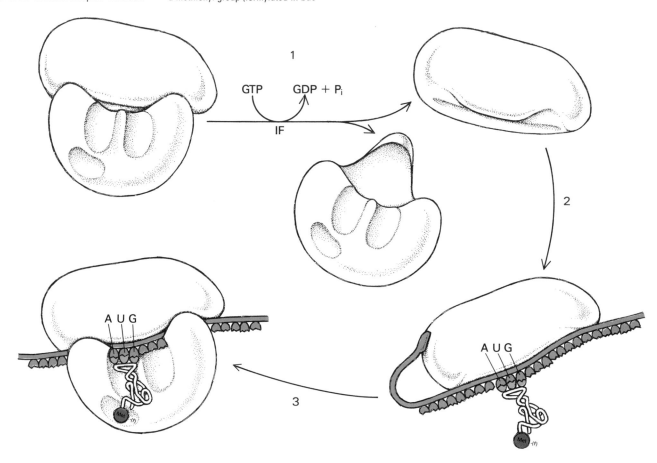

of the finished polypeptide, of its tRNA, and finally of the mRNA.

Initiation begins with the dissociation of such a free ribosome into its two subunits, which is followed by the binding of an mRNA and an aminoacyl-tRNA to the small subunit. This process, which also costs one GTP, requires very accurate matching, since it determines the reading frame. Have the fit shifted by as little as one base, and an entirely different message will be read. For a correct start, the mRNA must be bound to the small ribosomal subunit in such a way as to position an initiator AUG or GUG codon above the subunit's part of the P site. It is guided in this by its 5′ leader sequence, which recognizes special binding sites on the small subunit. The tRNA that combines with the initiator codon in this complex is different from all other tRNAs, and especially from the tRNA^{met} and tRNA^{val} that recognize AUG and GUG, respectively, when these codons appear inside a message: it has the unique property of joining with a small ribosomal subunit without participation of the large subunit. This initiator tRNA bears a methionyl group, which, in bacteria but not in eukaryotes, is N-formylated. All polypeptides, therefore, are born with an N-terminal methionyl residue, which, however, is often removed upon

Concept of reading frame. The figure shows how a message can be read in three distinct frames, to give two distinct peptide chains and an abortive runt.

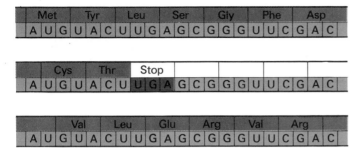

Met	Tyr	Leu	Ser	Gly	Phe	Asp
A U G	U A C	U U G	A G C	G G U	U C G	A C

Cys	Thr	Stop				
A U G U A C	U U G	A G C G G G U U C G A C

Val	Leu	Glu	Arg	Val	Arg	
A U G U A C U U G A G C G G G U U C G A C

subsequent processing. In bacteria, the formyl group is removed, but the N-terminal methionine is frequently conserved. The triple complex of small subunit, mRNA, and (formyl)methionyl-tRNA is called the initiation complex. Once it is formed, it binds a large subunit; elongation can start.

Such, greatly simplified, is the sequence of events in protein synthesis. On the whole, when you come to think of all the different recognition marks, binding sites, and catalytic activities needed by a ribosome for the correct performance of its duty, the complexity of structure of this "perfect robot" becomes understandable: even half a million atoms may not seem excessive. In spite of this complexity, the system has evolved to a considerable extent. The cytosolic ribosomes of eukaryotic cells are larger than those of bacteria, and many of their recognition sites and ancillary factors are different. This is very fortunate for us, since most antibiotics (though not penicillin, which inhibits the formation of the bacterial cell wall, Chapter 2) are inhibitors of protein synthesis. Thanks to the differences between the bacterial systems and our own, the drugs thus act selectively on the bacteria without harming us. The differences between eukaryotic and prokaryotic protein synthesis do, however, complicate the job of those genetic engineers who try to get a eukaryotic message (for instance, that for human insulin) translated in bacteria (see Chapter 18). They have to tamper with the leader part of the message to make it

readable by the bacterial machinery. Finally, you will no doubt remember that these differences have provided some of the most convincing arguments in support of the bacterial origin of mitochondria and chloroplasts. The ribosomes of these organelles are of bacterial type.

The Homing of Newly Made Proteins

With the exception of the rare proteins that are made in mitochondria or chloroplasts, all the proteins of a cell are initiated in the cytosol. Eventually, however, each of them finds its way to a specific intracellular compartment. The question is: How? We can obtain revealing insights into this question by taking a closer look at the "crawling polysomes," those ghostly centipedes that we discerned at the back of the tufts of growing secretory polypeptides on our tour of the rough ER (Chapter 6). We see them much better now as sinuous strings of a dozen or more ribosomes nailed to the ER membranes. We also know what they are doing, even though much of it is hidden. They are busily translating a tape of mRNA and delivering the polypeptide products of this process into the ER lumen. Let us focus our attention on the rapidly lengthening head of the centipede—the 5′ end of the mRNA, recognizable by its typical cap—as it emerges from the foremost ribosome of the string. For a while it stretches out freely, swinging and swaying with the cytosolic currents. Soon, however, a passing small ribosomal subunit gets caught by the mRNA's leader sequence; an initiation complex is rapidly built and joined by a large subunit; and translation of the message once again is set in motion. Now is when our sharpest scrutiny is needed. We have only a few seconds to watch a very remarkable event, the construction of a molecular secretory apparatus.

At first, things proceed as on free cytosolic ribosomes: the N-terminal tip of the nascent polypeptide starts growing out of the ribosome as a silky, undulating thread. But then, by the time the thread has become from twenty to

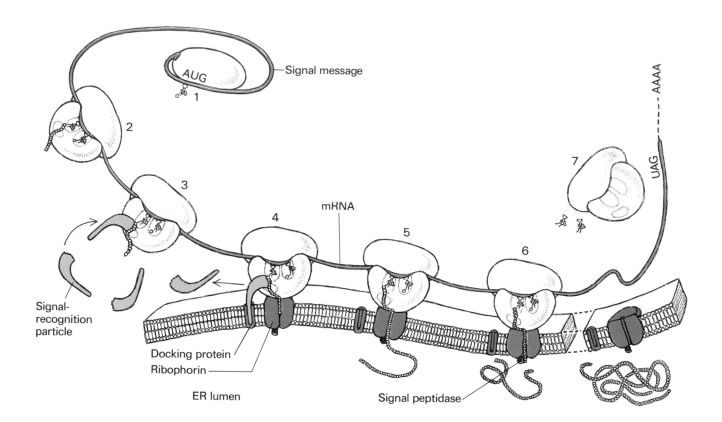

Signal message

1

AUG

2

3

mRNA

4

5

6

7

AAAA

UAG

Signal-
recognition
particle

Docking protein

Ribophorin

ER lumen

Signal peptidase

Cotranslational transfer of secretory proteins into the ER lumen, as determined by an N-terminal signal sequence.

1. Formation of an intiation complex in the cytosol.

2. Translation of the signal message has started.

3. Emerging signal peptide binds a specific signal-recognition particle (SRP) present in the cytosol. Protein synthesis is arrested.

4. Thanks to the affinity of the signal-recognition particle for an ER-linked docking protein (SRP receptor), the ribosome becomes moored to ribosome-binding proteins (ribophorins) associated with ER membranes. The SRP detaches, polypeptide assembly resumes, and the growing chain is discharged into the ER lumen.

5. The polypeptide has grown to the point that the signal peptide has been entirely extruded.

6. The signal peptide is split off by a signal peptidase.

7. Protein synthesis is completed. The ribosome detaches from the membrane.

thirty amino-acid residues long, it suddenly becomes entangled with a voluminous molecular complex, much in the manner of a surface receptor catching a ligand, and with equally dramatic consequences. For the caught complex actually stretches up to the ribosome itself, embracing it in such a way that chain elongation is arrested. N-terminal amino-acid sequences that have the ability to trap this jamming device are called signal sequences because of their directional role. The complex itself, which consists of protein and of a small 7S RNA component, is designated signal-recognition particle (SRP). A number of signaling sequences are known. They have some common features—in particular, a large abundance of hydrophobic residues. They are found at the 5′ end of all nascent secretory polypeptides. By binding the SRP complex, they activate an emergency brake that opposes the completion and delivery of such export materials in the cytosol. This state of affairs persists until the SRP-strangled ribosome bumps into a receptor, known as the docking protein, associated with the cytosolic face of ER membranes. Thanks to the specific interaction between SRP and docking protein, the ribosome becomes firmly anchored to the

ER by means of special integral membrane proteins, known as ribophorins. Its duty done, the SRP relaxes its paralyzing hold on the ribosome and falls off. Assembly of the polypeptide resumes; its product is inserted through the membrane and discharged into the ER lumen with the help of the ribophorins and, perhaps, other components, which make up the channel through which we saw the polypeptides grow into the ER lumen. Once this machinery is installed and molecular secretion is proceeding smoothly, the signal peptide is clipped off by a signal peptidase.

As a new ribosome is added to the head of the centipede by this mechanism, another one detaches from the centipede's tail, its link to the membrane severed by the detachment of the polypeptide it has just finished making. And so polysomes "crawl" on the surface of ER membranes, acquiring new ribosomes at the 5′ end and losing old ones at the 3′ end of their connecting mRNA. Occasionally, a brand-new mRNA becomes similarly involved, to start a new polysome creeping on the membrane in compensation for the inevitable wear-and-tear. These are the phenomena that take place on the cytosolic face of the membrane to produce the silky growth that made such an impression on us when we first set eyes on it.

This kind of cotranslational transfer—that is, transfer across a membrane while the message is being translated—seems to be of very general significance for export proteins, even in bacteria, where secreted exoenzymes are made by polysomes attached to the plasma membrane. It also plays an important role in the positioning of ribosomes that make integral membrane proteins for the ER and related systems, including the plasma membrane. As we saw in Chapter 6, many of these proteins, though not all, are made by ER-bound ribosomes. At first, things start very much as they do for secretory proteins, except that the signal piece is not always cut off but may remain instead as a hydrophobic anchor embedded in the lipid bilayer. If so, and if cotranslational transfer otherwise proceeds normally, a membrane protein oriented toward the ER lumen will be produced. In many cases, translocation of the growing polypeptide is blocked at some stage,

as it passes through the membrane, by a specific internal amino acid sequence that acts as a stop-transfer, or halt-transfer, signal. It is often a long hydrophobic sequence that gets caught in the lipid bilayer. Chain elongation is, however, not arrested by this jamming of the delivery process; as it continues, the ribosome is progressively pushed away from the membrane with the electrostatic help of a cluster of basic residues that often follows the hydrophobic sequence. As a result of this detachment, the ribosome completes the C-terminal end of the polypeptide in the cytosol. A transmembrane protein, with its N-terminal end exposed in the ER lumen and its C-terminal end facing the cytosol, is generated in this manner. More complex mechanisms of the same general nature control the synthesis of membrane proteins with a more involved topology. Some membrane proteins that face the cytosol are made by free polysomes and subsequently become attached to the membrane by some exposed hydrophobic parts.

Much less is known about the homing of other segregated proteins. Some cell explorers claim to have seen polysomes crouching on the surface of mitochondria and other membrane-limited organelles that contain resident proteins. Most, however, have looked for such bound polysomes in vain, finding instead that the proteins of these organelles (except for the few that are assembled internally) are made by free polysomes, released into the cytosol, and transferred posttranslationally to their final abode. How this occurs is still poorly understood. One assumes, by extension of the signal hypothesis, that the proteins must include in their structure some peptide sequence or other molecular configuration that specifically recognizes, and associates with, the membrane of their host organelle. But then something must occur to force the whole bulky molecule across the membrane, and here is where imagination becomes stretched beyond the boundaries of current preconceptions. Participation of a removable signal-type sequence has been detected in some instances, though not in others. In any case, it is difficult to see how complete transfer of the molecule across the lipid bilayer of one or, more often, two or even three (chloroplast thylakoids) membranes can be accomplished by such simple means. In cotranslational transfer one can at least visualize the GTP-fueled ribosome as forcing the growing polypeptide through the membrane. No such push is available posttranslationally, and one tends to think instead in terms of an internally generated pull acting on the membrane-linked protein. Alternatively, there could be an energy-fueled protein-translocating system in the membranes. Unfortunately, explorers have not yet devised means for reaching inside a mitochondrion or chloroplast to watch what is going on on the inner face of the membranes of these organelles.

The Life and Death of Proteins

For many proteins, getting "signaled" to their location is only the beginning of a long story. We witnessed some of this on our return from the recesses of the ER through the Golgi apparatus to the cell surface. The secretory proteins that were floating downstream with us were undergoing all sorts of changes, including glycosylation, association with lipids, closure of disulfide bonds, proteolytic clipping, and paring and remodeling of carbohydrate side chains. Actually, this processing continues beyond the cell surface up to the final change that announces destruction, most often through endocytic uptake and lysosomal digestion.

Take insulin, for example, the hormone whose deficiency causes diabetes. It is manufactured by a special cell, called β-cell, that occupies small regions, called islets of Langerhans (Latin *insula*, islet), in the pancreas, a digestive gland associated with the gastrointestinal tract. In its active form, insulin consists of two short polypeptide chains, the A chain with 21 amino acids and the B chain with 30 amino acids, linked by two disulfide bridges. It is, however, made as a much longer single chain called preproinsulin. Ribosomes starting to make this molecule in the pancreatic β-cell are directed to ER membranes by an N-terminal signal sequence, which is cut off cotranslationally. This removes the "pre" part of the molecule, leaving proinsulin as the product delivered into the ER lumen. The next step is oxidative closure of the disulfide

Maturation of insulin. The total message corresponds to preproinsulin, which is converted into proinsulin upon removal of the N-terminal signal peptide. While traveling through the ER toward the Golgi, proinsulin is folded and closed in this conformation by disulfide bonds. Later in its transit through the secretory apparatus, proinsulin is hydrolyzed by a specific endopeptidase that removes the C peptide, leaving mature insulin, the secretory and active form of the hormone, to be discharged into the bloodstream by exocytosis.

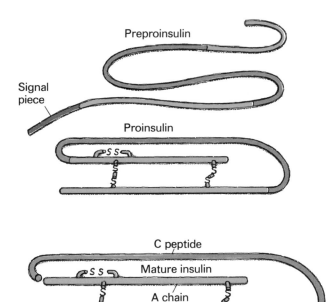

bonds. Then an internal sequence of 33 amino acids (C peptide) connecting the C-terminal A chain to the N-terminal B chain is excised by proteolysis, removing the "pro" part of the molecule.

The mature insulin is stored in tightly packed secretion granules, which, upon solicitation of the cell by an excess of sugar in the blood, discharge their contents exocytically into the bloodstream. The hormone now circulates, remaining in the blood until it happens to meet a free receptor on the surface of some cell. Binding of insulin to its receptor triggers a variety of intracellular events that result, among other effects, in an increased uptake of glucose, thus correcting the disturbance that stimulated insulin secretion in the first place. Providing a precaution against overreaction, attachment of insulin to its receptor also stimulates certain braking effects, among them endocytic uptake and digestion of the insulin in the lysosomes. All is over for this molecule. Joining with its receptor was both a fulfillment—the crowning event in its molecular life—and a death warrant, ordered by the needs of homeostatic regulation.

The fate and possible function of the C peptide are not known. But cases do exist where two or more active agents arise from a single polypeptide chain. A striking example is found in the anterior pituitary, a small, but very important, endocrine gland situated in the middle of the brain. Certain cells of this gland, called melanocorticotropic, make a signal-directed secretory polypeptide of 134 amino acids that is split into two hormones— ACTH (39 amino acids), which stimulates the adrenal cortex, and β-LPH, which acts on lipid metabolism—and a small connecting pentapeptide. Further carving of these molecules yields α-MSH (the first 13 amino acids of ACTH), β-MSH (amino acids 41 through 58 of β-LPH), and β-endorphin (amino acids 61 through 90 of β-LPH), of which the N-terminal pentapeptide forms enkephalin. The MSH factors, or melanotropic hormones, affect pigmented cells; they stimulate the changes in chromatophore color, which, as we have seen, depend on microtubule function (Chapter 12). β-Endorphin and enkephalin belong to a special group of brain peptides that mimic the effects of morphine and to which it is said we owe some of our more pleasurable sensations. These are not isolated examples. Hardly suspected a few years ago, the production of all sorts of biologically active peptides through the intracellular processing of larger precursors has now become one of the most exciting areas of research, especially in neurobiology and psychochemistry.

For a number of secretory proteins, the main carving-up occurs extracellularly: the proteins are secreted as inactive precursors, which subsequently are activated by proteolytic cutting when and where needed. A number of

Six hormones in one polypeptide. The diagram shows how different ways of carving the melanocorticotropic hormone of the anterior pituitary gland can produce as many as six different active products.

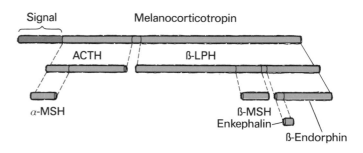

Signal   Melanocorticotropin

ACTH   β-LPH

α-MSH   β-MSH
                         Enkephalin
                                        β-Endorphin

digestive enzymes belong to this group; they are discharged as harmless zymogens and are unleashed only after reaching the stomach or the gut. Many proteins carried by the bloodstream are similarly manacled, ready to spring open upon the right proteolytic attack to produce the powerful—but deadly, if not properly controlled— factors that cause blood coagulation, clot dissolution, enlarging or narrowing of small blood vessels, mobilization of lymphocytes and other members of the immunological police, killing of foreign invaders. As we saw in Chapter 6, some of these events are controlled by a cascade of proteolytic attacks.

The membrane proteins made in the ER also go through many modifications, starting frequently with the cotranslational clipping of a signal piece. After completion, they start drifting by lateral diffusion in the plane of the lipid bilayer. During this voyage, they may undergo various kinds of processing on both sides of the membrane. A surface receptor protein, for instance, may acquire its oligosaccharide "hairs" in the lumen and part of its triggering mechanism in the cytosol. The itinerary and terminal destination of each protein are governed by the attachments they establish with their neighbors, with cytosolic components, and with luminal (or, eventually, external) ligands. Some proteins stay anchored in the ER, sometimes even, like the ribophorins, in the rough-surfaced parts of the ER. Others move on into the Golgi and become part of that structure. Yet others, many of them heavily glycosylated, get carried even further, either to lysosomes (with their mannose-6-phosphate-bearing ligands) or to the cell surface by way of secretory vesicles or granules. Once in the plasma membrane, they may become clustered into patches of different composition, participate in various forms of endocytic uptake, and become caught in the complex circuits described in Chapter 6. Eventually they end up in lysosomes after being segregated by autophagy.

Other intracellular proteins experience equally eventful, but different, adventures. They bind metals, acquire cofactors, assemble into multienzyme units, suffer the periodic addition or removal of groupings, such as phosphate or AMP, that affect their catalytic properties, until they, too, go the way of all mortal things.

In short, proteins have a life history. Between their ribosomal cradle and their lysosomal (or some other) grave, they go through a number of different stages, some of them associated with different locations. The whole of this molecular saga, as has been pointed out several times, is programmed in what is a protein's sole endowment at birth, its primary structure, written as a specific amino-acid sequence. Such linear sequences will then, either spontaneously or with the collaboration of some of the components offered by the complex intracellular environment, fold, unfold, and refold into defined three-dimensional configurations, which will themselves exhibit catalytic properties, display binding sites, offer targets to other enzymes, or do any of the thousands of other things proteins do in their lifetime.

Even death is included in this program, not in terms of a rigidly determined life-span, but rather in terms of a probability. The death of proteins is a stochastic process, and their life-span is a statistical notion generally defined by their half-life—that is, the time needed for half the molecules of a given species that have been synthesized at the same time to be destroyed. Half-lives are measured with the help of labeled precursors.

Just imagine that while we are visiting this cell it is exposed for a short time to a mixture of amino acids

The concept of turnover. At time zero, a radioactive amino acid has been provided for a brief length of time (pulse) during which radioactive proteins are synthesized. Proteins made after that to replace those destroyed are again nonradioactive. With a constant protein pool (steady state), the radioactivity in the pool falls in exponential fashion. The half-life of the protein is the time needed for the specific radioactivity of the protein to decrease by one-half.

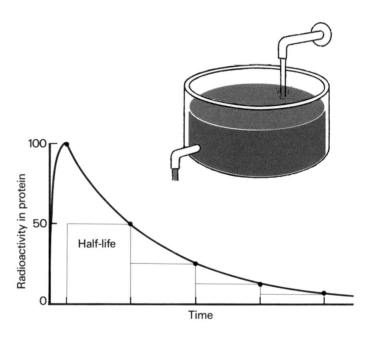

painted red. We then would be able to distinguish the proteins born during this brief period by their color and, therefore, to watch their disappearance and to measure their half-lives. The beauty of this technique is that it hardly disturbs the cell, except for the possible effect of the paint. For the red proteins that disappear before us are being replaced by newly synthesized colorless ones. The total number of molecules does not change. The cell is in a steady state; we are watching turnover without interfering with it. In reality, of course, we do not use red amino acids, but radioactive ones, at low enough dosage to keep radiation damage to a minimum. When such experiments are done, it is found that the average half-life of proteins is of the order of two to three days, but it varies greatly from one protein species to another. Half-lives are measured in minutes for some proteins, in hours for others, in days or sometimes even in weeks or months for yet others.

Where Is the Boss?

Many of the events that make up the life history of individual proteins are themselves mediated, directly or indirectly, by other proteins. The outcome is a four-dimensional, multifactorial network of interactions, which may, in consideration of the overwhelmingly important role of proteins, be viewed as conditioning the life of the cell itself. Different cells are different essentially because they have a different complement of proteins. When such cells associate to form an organism, a network of networks of interactions is produced. In turn, as organisms join into groups, societies, ecosystems, up to the whole biosphere, networks of higher and higher order are built up, finally to embrace, in a single interwoven unit of incredible complexity, the thin pellicle of living matter that covers our planet.

Life, in all its forms and activities, is to a very large extent a manifestation of proteins and, more precisely, of certain associations of proteins capable of generating a self-stabilizing, self-regulating network of interactions adapted to certain environmental conditions—that is, a viable system. But the proteins themselves, as we have just found out, are but the expression of the mRNAs supplied to the ribosomes of each cell. Therefore, we may reformulate the statement above and say that life is a manifestation of mRNAs. In fact, this generalization is even more correct than the first one, if we make it include all RNAs, not just mRNAs. For the mute sequences that are not translated, and especially the tRNAs and rRNAs, also play key biological functions that are strictly determined, as are the messages, by specific base sequences.

And this then raises the ultimate question: Who or what specifies the RNAs? Who or what spells out the instructions? Who or what writes the scores that are so magnificently, but quite slavishly, executed by the cytoplasmic orchestras? To find the answers to these questions, we must penetrate the inner circle, the hidden center where each cell's repertoire of information is on file. We must enter the nucleus. There the decisions are made instant by instant, or rather seem to made—there really is no boss, as we shall see in Chapter 16—about what instructions to send out and in how many copies.

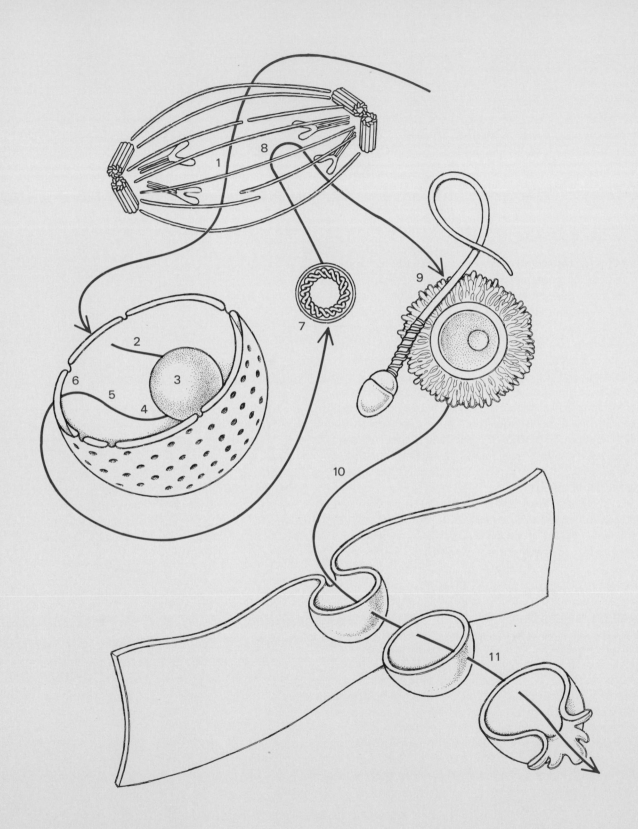

ITINERARY III | The Nucleus

1   Take advantage of mitotic division to enter newly forming daughter nucleus.

2   Watch chromosomes uncoil into immense lengths of DNA.

3   Visit nucleolus and observe synthesis of ribosomal RNA.

4   Move on to euchromatin area and inspect DNA transcription into messenger RNA.

5   Wait until a mitogenic stimulus initiates DNA replication and examine this process.

6   Remain in interphase nucleus to observe hidden movements of allegedly resting DNA.

7   Follow some viruses on their rounds and pause to reflect on a few problems, such as cancer, genetic engineering, evolution, and the origin of life.

8   Contemplate mitosis and separation of duplicated chromosomes.

9   Move briefly to a germ cell and take a look at meiosis.

10 Leave nuclear area and remain in the cytoplasm to watch the end of cell division.

11 Use plasma membrane to bud out of the cell and make final exit.

16 | Transcription and Editing of Genetic Messages

In Latin, *nucleus* is the diminutive of *nux*, nut. The Greek word *karyon*, which has lent its root to such terms as eukaryote, prokaryote, karyolysis, karyotype, heterokaryon, likewise means nut. The nucleus derives its name from its location: it sits in the center of the cell like a stone in a cherry. Also like a cherry stone, it keeps its precious kernel enclosed within a protective shell.

A Well-Defended Citadel

The nuclear shell or envelope is not hard and rigid; it is membranous. At first sight, you might well mistake it for a rough ER cisterna; it is made of the same tenuous material and is studded with typical bound polysomes. As a matter of fact, you would not be wrong, since the nuclear envelope is indeed part of the ER. But it is a very special part, huge and completely bent and fused around the nuclear contents to form two concentric membranous spheres separated by a narrow space some 10 to 15 nm wide. This space is a true cisternal space; it receives secretory material made by the bound polysomes and transfers it by means of connecting channels to the more conventional parts of the ER. The polysomes, however, occupy only the outer wall of this remodeled cisterna, the one facing the cytosol. The inner wall is smooth; there are no ribosomes inside nuclei.

The structure enclosed by this double-walled envelope is of considerable size. It has a diameter of 8 to 10 μm and occupies one-tenth or more of the cell volume. At our millionfold magnification, this makes for a roomy cham-

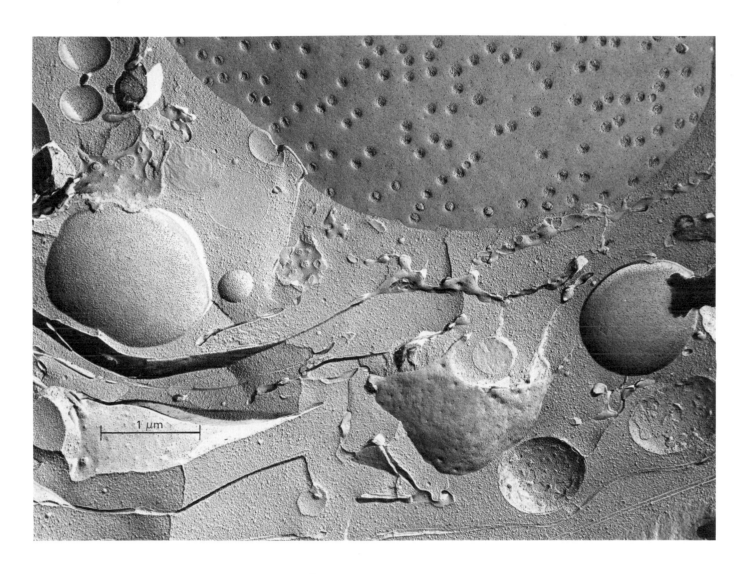

ber, some 25 to 30 feet in diameter. Its casing is flimsy, hardly more than a quarter of an inch thick (7 nm, real size), which is even thinner than the plasma membrane. But it is double and bolstered by a strong protein lining called the lamina. The nuclear envelope acts as an excellent barrier between the nucleus and the cytoplasm, forcing all exchanges to take place through special channels known as pores.

The imprint of the nuclear envelope and of its pores appears in the upper part of this freeze-etch electron micrograph of part of an onion (*A. cepa*) root-tip cell. Various fractured organelles can be seen in the cytoplasm.

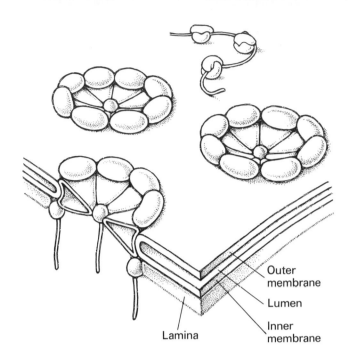

Schematic structure of the nuclear pore complex. The double-walled nuclear envelope, derived from the endoplasmic reticulum (notice the polysome on the cytoplasmic face), is pierced by thousands of round openings, or pores. Each pore is bordered by a continuous seal between the inner and outer membranes. The openings are largely closed by a proteinaceous complex made of eight conical flaps, surrounded by an inner and an outer ring (annulus) of eight globular subunits. The central hole is often stoppered by a plug. The pores are interconnected by a fibrillar web, called the lamina, apposed against the inner face of the envelope.

Labels: Outer membrane, Lumen, Inner membrane, Lamina

The nuclear pore is a proteinaceous structure with an eightfold radial symmetry inserted into a windowlike eyelet, or fenestration (Latin *fenestra*, window), bored through the two membranes of the envelope and sealed by their fusion. The outer diameter of the complex is about 120 nm, that of the hole itself, from 60 to 90 nm. But much of this is closed by a diaphragmlike structure consisting of eight conical flaps, leaving an opening of only about 10 nm, which is itself often stoppered by a plug. At our magnification, a pore would appear as a small, octagonal porthole with an overall diameter of 5 inches and a central aperture of a little less than half an inch. In our "average" mammalian cell, there may be as many as several thousand such pores distributed over the whole surface of the nuclear envelope. They give the envelope a sort of padded appearance and distinguish it clearly from ordinary ER cisternae. Judging from their structural intricacy, nuclear pores are not likely to be simple holes. They must be dynamic entities, actively mediating and regulating the transport of materials in and out of the nucleus. Unfortunately, we have as yet no inkling of how they operate. All we can do is to watch the result; a worthwhile experience in itself, which tells us a great deal about the nucleus and its functions even before we get inside.

Most striking is the steady outflow of RNA. Representing the principal product of nuclear industry, freshly made RNA threads are being extruded continually through every pore, at the rate of about 1 μm (more than 3 feet at our magnification) per pore per minute. These RNA threads come out in association with proteins, which presumably help them through the pores. Much of the RNA produced by the nucleus is rRNA, destined for the construction of the thousand-odd ribosomes that a cell must make every minute to compensate for the ineluctable attrition that affects ribosomes as it does everything else in the cell. Replacement of tRNA likewise makes a continual demand on nuclear manufacturing activity, as does the production of the small sRNAs that participate, for instance, in protein signaling (Chapter 15). Finally, a small but crucial part of the RNA that flows out through the nuclear pores is mRNA; it contains the bulk of the ordinances issued by the nucleus. In addition to RNA, a few other important molecules are made in the nucleus and delivered

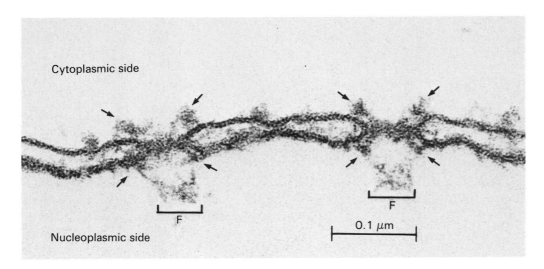

Cytoplasmic side

Nucleoplasmic side

F

F

0.1 μm

Electron micrograph of a cross-section through the nuclear envelope of a *Xenopus laevis* oocyte. Note the double membrane with its cisternal lumen and the lamina bolstering the membrane on the nucleoplasmic side. Two pores are seen: the arrows point to annular granules. Filament bundles (F) are attached to the nucleoplasmic side of the pore complex.

to the cytoplasm through the pores, among them NAD.

Flowing in the reverse direction through the nuclear pores are the activated building blocks of RNA synthesis. Except possibly for some glycolytic activity provided by cytosolic enzymes seeping through the pores, the nucleus has no autonomous supply of energy. Its biosynthetic role is restricted to assembly. It relies entirely on the cytoplasm for both raw materials and power, supplied jointly in the form of the four key nucleoside triphosphates: ATP, GTP, CTP, and UTP (or nucleoside mono- or diphosphates and a corresponding excess of ATP for their phosphorylation). When called upon to replicate its DNA, the nucleus likewise receives its energized substrates from the cytoplasm in the form of the appropriate deoxynucleoside triphosphates—dATP, dGTP, dCTP, and dTTP—or of their precursors. These supplies are complemented by a substantial additional amount of ATP that is used for purely energetic purposes and returned to the cytoplasm as ADP or AMP for recharging. The nuclear pores also allow a considerable inward flow of proteins, comprising both newly made nuclear proteins, which are all synthesized by free cytosolic polysomes, and cytoplasmic proteins that come to fetch RNA out of the nucleus. Finally, if you look carefully enough, you will distinguish in the inward stream a number of rare molecules, some of them of protein nature, others as yet unidentified, that serve to carry instructions from the cytoplasm to the nucleus.

Nuclear pores thus let many things into and out of the nucleus, though not, however, the members of our group. Try as we might, we could not possibly squeeze through one of those narrow portholes. To enter forcibly, we would have to rip apart the fragile fabric of the envelope, possibly causing irreparable harm to the delicate balance between cytoplasm and nucleus. If we wait long enough, however, our opportunity may yet come, when the cell that we are visiting enters into division. At that time, the nuclear envelope fragments into vesicles; the barrier between cytoplasm and nucleus disintegrates. All we have to do is ensconce ourselves near one of the two clusters of chromosomes that separate during mitosis (Chapter 19), and stay there. Soon a new envelope will form around us, and we will find ourselves inside the nucleus of one of the daughter cells.

Inside the Nucleus

Witnessing the reconstruction of a nucleus from the inside is a stifling experience. While you wait in the midst of the chromosomes, large expanses of endoplasmic reticulum start closing in from all sides, pressing more and more, until by the time they all come together and fuse into a single hermetic hull, you are left with hardly any space in which to move. At the same time, your main landmark disintegrates before your eyes. The chromosomes, those solid pillars under which you had taken shelter, fray progressively into interminable lengths of a ropelike material, snarling you in thousands of tangled coils. It is enough to drive the hardiest explorer into a claustrophobic panic.

There is, however, no need to yield to such an unreasonable fear; surely there will be a way out. In the meantime, let us take stock of our surroundings and see what more we can learn about the cell now that we are inside its most jealously guarded citadel. The ropes that are all around us are chromatin fibers, deriving their name, as do the chromosomes from which they originate, from the Greek word for color. Not that they are themselves colored; they are snow white, in fact. But they avidly bind a number of basic dyes. Thanks to this basophilia, they appear brightly colored in appropriately stained preparations, and were named after this property by the investigators who first saw them about a century ago. To be sure, those early explorers did not see the actual fibers. They simply recognized chromatin as the main nuclear substance and chromosomes as rod-shaped condensations arising from it during cell division.

We don't suffer the same limitations and readily see that chromatin is made of fibers some 25 nm thick. In a human cell nucleus, there are 46 such fibers, one per chromosome, varying in length between about 0.25 and 2 mm. This may not look like much until you scale it up to human size. Magnified a millionfold, it amounts to some 30 miles of hefty rope, about 1 inch thick, packed in a 30-foot ball.

Yet, this is not all. The chromatin ropes, you will notice, are actually made of a thinner, 10-nm-thick strand, coiled into some sort of solenoid, much like a telephone-receiver cord. When unrolled, this strand looks like a string of beads. Each repeating unit of this structure is known as a chromatosome; the beads are called nucleosomes. There are six nucleosomes per turn of the spiral, which means that if we completely uncoiled the ropes their length would increase sixfold, up to a total of almost 12 inches in real size for the 46 of them, or 180 miles, when magnified a millionfold.

The string of beads is itself made of even thinner twine, only 2 nm thick, and of small bundles, which make up the cores of the beads. The twine runs continuously from core to core, taking about two turns around each before moving on to the next. Should we unwind the twine from its spools, another sixfold lengthening would ensue, up to a total of about 6 feet per nucleus. Scaled up a millionfold, to the point where the nucleus forms a 30-foot sphere around us, this twine would look like ordinary string, a little less than one-tenth of an inch thick, and stretching, when all 46 pieces are strung together, over more than 1,100 miles, the distance from New York to Kansas City.

We can finish this molecular dissection by adjusting our glasses to their highest resolving power. We now see that the twine is itself double stranded and helically coiled. It is none other, in fact, than the celebrated double helix, the ultimate molecule of life, the symbol of modern biology, DNA. As to the small bundles that make up the cores of the nucleosomes, they consist of a special class of basic proteins named histones. Two molecules each of four different types of histones, designated H2A, H2B, H3, and H4, make up the core. Another histone, known as H1, is associated with the short strand of DNA that connects one bead to another. H1 molecules probably join in some sort of axial shaft when a string of nucleosomes coils into a chromatin fiber.

Most of the nuclear space is taken up by these endless coils of chromatin, which, densely packed in some places, partly unrolled in others, create a picture of utter disorder and confusion. Much that same impression might be left by the inside of a telephone switching center or a computer until we look more closely at the wiring pattern. Indeed, if we choose any part of a chromatin coil and

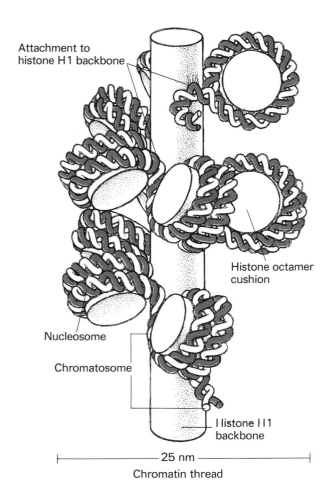

Attachment to
histone H1 backbone

Histone octamer
cushion

Nucleosome

Chromatosome

Histone H1
backbone

├──────── 25 nm ────────┤
Chromatin thread

Fine structure of chromatin thread. This drawing illustrates the solenoid model, with a string of nucleosomes coiled helically, with a pitch of six nucleosomes per turn, around an axial structure believed to consist of aggregated histone H1 molecules, possibly associated with other proteins. Each nucleosome has an octameric protein core, roughly cylindrical in shape, made of two molecules each of histones H2A, H2B, H3, and H4, around which are spooled 1.75 turns of double-stranded DNA (about 50 nm). DNA runs continuously from nucleosome to nucleosome. Short linker segments of DNA between nucleosomes are attachment points to histone H1. They vary somewhat in length, depending on cell type and species. The whole repeating unit (chromatosome) contains between 160 and 240 nucleotide pairs (54–82 nm).

follow it long enough, we will find that it leads us to a protein framework called the nuclear matrix. Almost impossible to discern in an intact nucleus, the matrix is readily seen after chromatin has been selectively removed by appropriate chemical or enzymic treatment. It then appears as a three-dimensional network that pervades the nucleus. It is likely that the elements of this network originate from the protein scaffolding that supported the chromosomes during mitosis and that they will participate in the rebuilding of this scaffolding at the beginning of the next mitosis. As we shall see, chromatin is anchored to the nuclear matrix in such a way as to divide each fiber into a succession of neatly separated loops, or domains. Behind the embroilment of tangled fibers, there is in fact a great deal of order.

A Bit of History

It is difficult for us, children of the Watson-Crick era, even to visualize that not so long ago most people believed DNA to be too simple a substance to have any but an accessory role in heredity. Its association with the chromosomes was, of course, known, and so was the role of the chromosomes as bearers of the genes. But only the chromosomal proteins, perhaps in the form of "nucleoproteins," were considered to have the chemical complexity and inexhaustible possibility of variation required for a genetic material.

To understand this, we must travel back more than 100 years to two events that jointly launched biology toward its greatest conquest to date, but whose fundamental interconnection remained unsuspected for a long time. One took place in the garden of an Austrian monastery, where a monk named Gregor Mendel patiently counted the types of offspring that he obtained by crossing different varieties of peas. Out of this work came the "laws of Mendel," first published in an obscure local journal in 1866, but largely ignored until the turn of the century, when they were rediscovered. From them arose the science of genetics, a term derived, like genesis, from the Greek verb *gennân*, to beget, to generate. The essence of

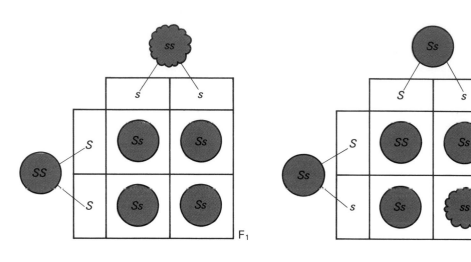

One of Mendel's crossing experiments. Cross-fertilization of two varieties of peas differing by shape of seeds gives first-generation (F_1) hybrids that produce only smooth seeds. Self-fertilization of the F_1 hybrids gives second-generation (F_2) hybrids, of which three-quarters produce smooth seeds and one-quarter wrinkled seeds. The diagram shows how these results can be interpreted by assuming that the shape of a seed is determined by a pair of "units" (alleles), each of which is either dominant (S) or recessive (s), and that allelic units are segregated in germ cells and joined at random upon fertilization.

Mendel's discovery, expressed in modern terms, is that discrete hereditary traits—as defined by the color, shape, texture, or other morphologically recognizable property of certain organs that is transmitted from one generation to the next—are determined by units, the genes, that always go in pairs, the alleles (Greek *allos*, other), of which one is provided by the male, the other by the female parent. The character of the offspring depends on what alleles, dominant or recessive, it receives from its two parents. Individuals with the same two alleles are designated homozygotes (Greek *homoios*, same; *zygos*, pair, couple). They are what we would call "pure race" (for that particular trait) and upon inbreeding continue, generation after generation, to express the character determined by their identical alleles. Heterozygotes express the dominant character, but only three-quarters of their offspring do so; one-quarter express the recessive character. Why this must be so is obvious. If S is the dominant and s the recessive allele, mating two heterozygotes, Ss, with random selection of the allele donated by each parent, will produce, in equal proportion, the combinations SS, Ss, and sS, all of which express the dominant character, and ss, which does not.

This is what Mendel found more than a century ago, except, of course, that he found it the other way round. He discovered the 75/25 distribution empirically and deduced the allele model from it. Inevitably, numerous exceptions and complications to this simple model were recognized subsequently. The most important one in relation to the present overview is the presence of linkage, or coupling, between genes, indicating that genes are not independent units, but are physically associated as sets.

The other major event that happened in those early days took place in 1869 in the laboratory of the German physiological chemist Felix Hoppe-Seyler, at the University of Tübingen, where a young Swiss physician, Friedrich Miescher, isolated a previously unknown substance from nuclei. His starting material consisted of pus cells isolated from septic bandages. In addition to carbon, oxygen, hydrogen, and nitrogen—the four elements commonly found in proteins and other biological materials—the new substance was found to contain large amounts of phosphorus. Miescher named his discovery nuclein, a word that was later changed to nucleic acid after the strongly acidic character of the substance was recognized. It took more than sixty years and the efforts of many celebrated chemists—in particular, the German Albrecht Kossel and the American Phoebus Levene—to unravel the structure of the new substance and of its components. In the meantime, the presence of a similar material in the cytoplasm was recognized. The two nucleic acids were first known as thymonucleic (from thymus, one of the richest sources of animal DNA) and zymonucleic (from the Greek word for yeast, an early source of RNA). Later, when the nature of their respective constituent pentoses was uncovered, they were renamed deoxyribonucleic and ribonucleic acids.

What brought Mendel's units and Miescher's nuclein together was the work of the cytologists who discovered

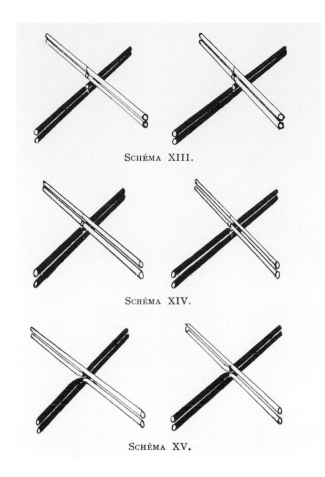

SCHÉMA XIII.

SCHÉMA XIV.

SCHÉMA XV.

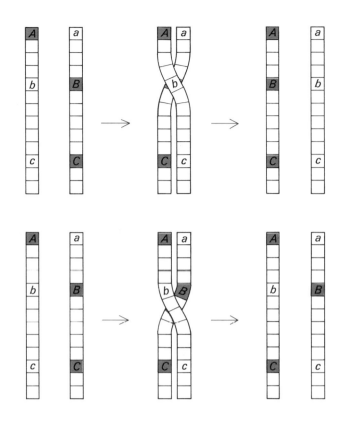

Crossing-over, or chiasmatypie. The drawing by Frans Janssens reproduced here was originally published in *La Cellule* in 1909. It illustrates the author's interpretation of the chi figure formed by two homologous pairs (dyads) of duplicated chromosomes (chromatids). After one chromatid of each pair has been broken at the point of overlap, the segments recombine crossways (for a description of meiosis, see Chapter 19).

The genetic implication of Janssens's theory of chiasmatypie, according to Thomas H. Morgan. Genes situated on distinct homologous chromosomes become recombined by crossing-over. In both cases, dominant allele *C* recombines with *A*, whereas *B*, which is situated closer to the *A* locus, recombines with *A* only in the first case. Thus, the probability of two genes situated on distinct homologous chromosomes recombining on the same chromosome by crossing-over is proportional to the distance separating the loci that they occupy on the chromosome. Hence, the measurement of recombination frequencies can be used for chromosome mapping.

the chromosomes and described the behavior of these structures in mitotic division and, especially, their numerical halving in meiosis. As we will see in Chapter 19, chromosomes go in pairs, as do the postulated alleles. They duplicate at every cell division to provide identical sets for the daughter cells, as should bearers of heredity. And, during maturation of the germ cells by meiotic division, only one of each pair is retained in either spermatozoa or ovules, so that, upon fertilization, fusion of the male and female nuclei recreates a full set of chromosomes in which every pair has one member contributed by each parent, which is exactly what alleles are expected to do.

Furthermore, the size and shape of chromosomes are consistent with the concept of a linear association of many genes, indicated by the linkage phenomenon. This became even more evident after a Belgian biologist, Frans Janssens, observed that during meiosis paired chromosomes often "cross over" each other and exchange large segments before separating again. This process, called chiasmatypie by Janssens (Greek *chiasma*, X-shaped cross), and later renamed crossing-over, provided an explanation for the puzzling fact that linkage is not an all-or-none phenomenon, but a graded one, depending on the genes involved. One can readily imagine that the

Caption on left side, then the map image on right.Partial genetic map of the four chromosomes of the fruitfly *Drosophila melanogaster,* as charted by the Morgan school. Genes are named according to known mutant character. The numbers at the left are distances inferred from recombination frequencies.

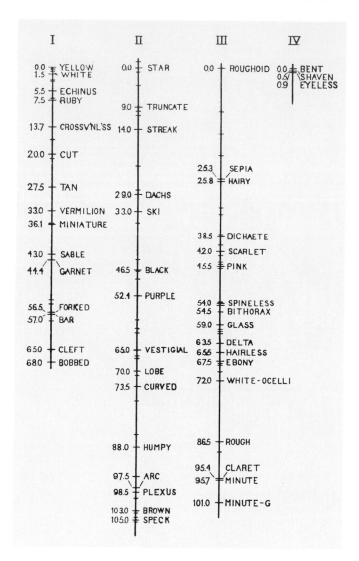

probability of two linked genes becoming unlinked by crossing-over must increase with the distance separating them along the length of the chromosome. Further details on crossing-over will be given in Chapters 18 and 19.

The major architect of the grandiose synthesis between genetics and cytology was Thomas Hunt Morgan, an American scientist who turned the quantitative measurement of the tightness of linkage between two genes into a method for estimating the distance that separates them on their bearer chromosome. Using this technique of genetic mapping, Morgan and his co-workers succeeded, between the years 1910 and 1922, in locating hundreds of genes on the four chromosomes of the small fruit fly *Drosophila melanogaster*. Later, these maps were given a true physical basis through correlation with the fine structure of the giant chromosomes that are in the salivary-gland cells of the fruit fly. Genes could thereby actually be localized to transverse bands on these chromosomes. Real flesh was put on Mendel's "units," but the nature of that flesh remained a puzzle.

Here, again, cytology provided the necessary bridge, but this time with chemistry. By using various stains, enzymic attacks, and other cytochemical means, investigators attempted to characterize chemically the structures that they saw through the microscope. This search demonstrated that the chromosomes consist mainly of DNA and protein. But it took several decades before it was recognized that the DNA component, and not the protein, is the actual bearer of genetic information. The final turning point was provided by what has become one of the greatest classics of scientific literature—a paper published in 1944, with, as authors, three scientists of The Rockefeller Institute for Medical Research in New York, Oswald T. Avery and two young colleagues, Colin MacLeod and Maclyn McCarty. The origin of Avery's work was an observation made in 1928 by Fred Griffith, a medical officer of the British Ministry of Health. Experimenting with pneumococci, the agents of bacterial pneumonia, Griffith had found that mice inoculated with living bacteria of a nonvirulent strain (called R, for rough, because it forms rough-looking colonies) and with dead bacteria of the virulent S (smooth) strain would die of pneumonia and yield at autopsy a culture of live S-type pneumococci. This remarkable phenomenon, called bacterial transformation, was subsequently reproduced in vitro by treating R-type bacteria with extracts of S-type bacteria. It was known that S-type pneumococci owe their virulence to the possession of a specific capsular polysaccharide, which is absent in the R type. Transformation, however, did not represent restitution of the pathogenic polysaccharide itself, as virulence persisted through countless bacterial genera-

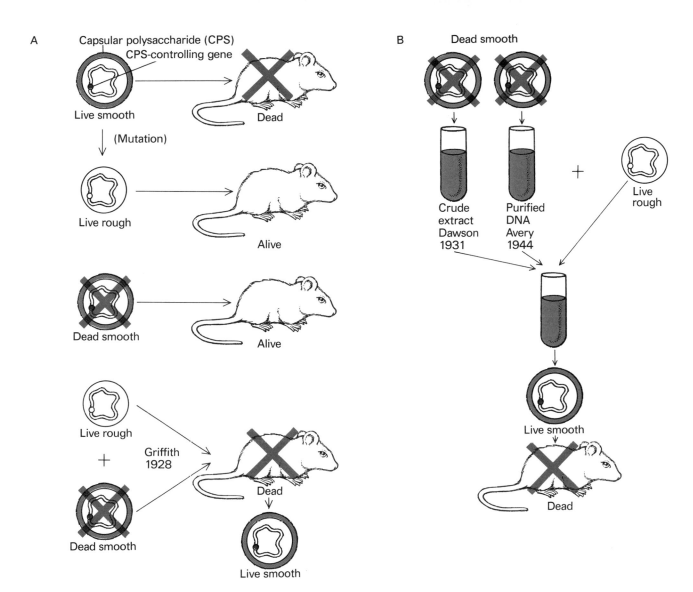

Bacterial transformation.

A. S pneumococci form smooth colonies, are surrounded by a polysaccharide capsule that protects them against destruction by the immune system, and are therefore highly virulent. R mutants form rough colonies, lack the gene for making the polysaccharide capsule, and are non-pathogenic. Heat-killed S bacteria are likewise nonpathogenic. However, as shown by Griffith in 1928, the inoculation of live R bacteria together with heat-killed S bacteria results in a lethal infection. Live S bacteria are recovered from the bodies of the infected animals.

B. Transformation of R into S pneumococci can also be effected in vitro with crude extracts of S bacteria, as first observed by Dawson in 1931. Purification of the transforming principle and its chemical identification as DNA were first reported in 1944 by Avery, Macleod, and McCarty.

tions. What was restituted or transmitted, therefore, was the ability to make this polysaccharide and to pass on the property to progeny. It had to be viewed as the insertion into the R bacteria of the corresponding gene, borrowed from the S type. Hence the immense interest in the chemical nature of the transforming factor. The famous 1944 paper was the climax of a long, patient attempt at purifying this factor. It established beyond reasonable doubt—except that of the critic willing to go to any length brandishing Avogadro's number—that the transforming factor was a DNA, not a protein.

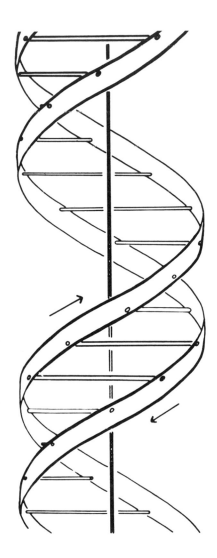

When Watson and Crick made their epoch-making proposal, the structure of DNA as a very long polynucleotide chain was common knowledge. In addition, they had available three types of facts:

Fact one, physical, was provided by X-ray diffraction data indicating a helical structure for the fibrous molecule, with parametric dimensions suggestive of double-strandedness rather than single-strandedness.

Fact two, chemical, came from the determinations of the Austrian-American biochemist Erwin Chargaff, who had found that, in samples of DNA of widely different origins, the adenine content was always equal, within limits of experimental error, to the thymine content, and the content of guanine to that of cytosine. Hence "Chargaff's rules": A = T; G = C.

Fact three, biological, was the necessity that the molecule be capable of inducing its own duplication.

These facts were known to the other runners in the race, who were all busily trying to build a molecular model of the genetic molecule. The flash of inspiration that put Watson and Crick on the correct track comes down to two words: *base-pairing*. You will remember that we have already encountered this phenomenon as playing a key role in the structure of RNA and in the translation of genetic messages. Historically, however, base-pairing was first discovered through a consideration of DNA, where it explained Chargaff's rules and at the same time suggested a mechanism for self-duplication. Each molecule would consist of two complementary strands, representing, so to speak, the same information in positive and negative. As in photography, the positive could serve as template for a new negative, the negative for a new positive, leading to the formation of two identical double-stranded molecules in which, incidentally, one strand would originate from the parent molecule and the other would be newly made (semiconservative replication). Once this intuitive concept had taken shape, it remained for the two investigators to find, by empirical model-building, a molecular arrangement that fitted base-pairing into a helical structure of appropriate dimensions without imposing any

By 1950, virtually everybody was convinced, and the race for DNA was in full swing. The winners: the American James Watson, 25, and the Englishman Francis Crick, 37, with a small, unobtrusive paper occupying less than two pages in the April 25, 1953, issue of the British magazine *Nature* and titled "A Structure for Deoxyribose Nucleic Acid." It introduced the double helix to the world.

undue strain on the distances and angles known to characterize each interatomic bond. The thrill and drama of this momentous discovery, and the climate of intense competitiveness and some of the less pleasant aspects of human behavior that surrounded it, are best appreciated in Jim Watson's own candid account, *The Double Helix*.

The Double Helix

The main structural characteristics of DNA have already been described. Like RNA, DNA is a polynucleotide, made by the linear association of 5'-mononucleotides linked by 3'-phosphodiester bridges. It differs chemically from RNA in two respects: (1) the pentose is 2-deoxyribose instead of ribose—that is, ribose without an oxygen in position 2; (2) one of the two pyrimidine bases is thymine instead of uracil—that is, uracil with an additional methyl group in position 5. These differences have little effect on a number of structural properties. In particular, they do not affect base-pairing, which is essentially similar in DNA and RNA because the A—U and A—T pairs have the same shape and strength. But certain chemical properties are greatly influenced by the differences between the two nucleic acids: RNA is much more sensitive to alkaline hydrolysis than is DNA; the latter, in turn, can be selectively deprived of its purine bases by mild acid treatment, a property that is made use of in the cytochemical Feulgen reaction, whereby DNA can be selectively stained in tissue sections.

The most important difference between DNA and RNA lies in their strandedness. Apart from some viral forms, all the RNA in the world is single stranded; it base-pairs mostly with itself, to form loops closed by short double-helical complementary segments. It is just the opposite for DNA. With the exception of a few small viruses, all the DNA in the world is made of two complementary antiparallel strands. It is double helical over its whole length and, in its most common form, known as B, appears as a regular, right-handed, spirally twisted thread with a diameter of 2 nm and a pitch of 10 nucleotide pairs per turn, each pair occupying an axial distance of 0.34

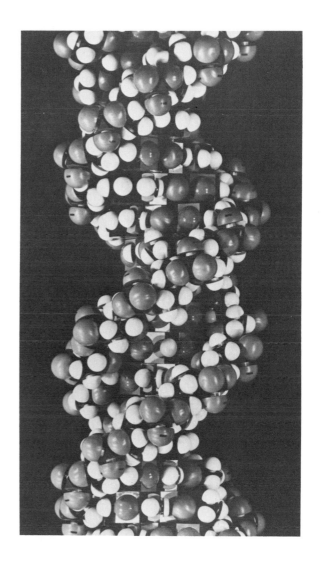

nm. Thus, 1 μm of DNA has about 300 turns, 3,000 nucleotide pairs, and a molecular mass of circa 2 million daltons (the average molecular mass of a nucleotide pair, sodium salt, is 660 daltons). These relationships have been verified directly, because techniques are available for visualizing DNA threads with the electron microscope. DNA can also adopt other shapes, including a newly discovered left-handed Z form, rare but possibly of biological significance.

Another difference between DNA and RNA lies in their length. In contrast with most RNAs, natural DNA threads are polycistronic. (The cistron is the molecular equivalent of a unit-gene, a sequence coding for a single polypeptide or RNA molecule.) Quite often, in the world of viruses and of prokaryotes, all of the genome is carried by a single, circular DNA thread. The length of this circle varies between 1.5 and 80 μm in viruses, and reaches 1 mm in bacteria. In eukaryotic cells, as we have seen, the DNA threads are considerably longer, up to nearly 10 cm. There is one such thread per chromosome, where it is coiled around small histone cushions to form a string of chromatosomes. The length of DNA per chromatosome varies somewhat from one species to another, with an average of about 200 base pairs (68 nm), of which a constant stretch is coiled toroidally around the histone cushion (nucleosome) and the remainder serves as internucleosomal link. As we have seen, the chromatosome string is itself twisted into a helical chromatin fiber. Depending on the state of the nucleus, this fiber is variously unrolled and dispersed. At the time of cell division, it coils and supercoils into the compact, rod-shaped structure of the chromosome. Unlike prokaryotic DNA, eukaryotic DNA is not circular. Those who like figures may be interested in learning that each of us is the possessor of more than 400 billion billion (4×10^{20}) nucleosomes and of about 20 billion miles of DNA—enough, if completely stretched, to run back and forth 100 times between the earth and the sun!

For obvious reasons, double-stranded DNA is more rigid than single-stranded RNA. Nevertheless, a DNA thread can assume all sorts of contorted shapes. The way it coils around nucleosomes and twists into chromosomes makes this abundantly clear. But these contortions generate strains and create topological problems that make even the most intricate combination of Möbius strips look like child's play. These problems are real, not just conceptual. They are encountered by the enzyme systems that transcribe and duplicate DNA and by the proteins that package it. They call for the participation of special enzymes for loosening structures, unwinding coils, and disentangling knots. We will run into several of these topoenzymes as we watch DNA actually imparting genetic information.

The stability of a double-stranded DNA thread depends on two types of forces: (1) the electrostatic hydrogen bonds that determine base-pairing; and (2) hydrophobic interactions between the planes of the stacked base pairs. Opposing these stabilizing forces are the electrostatic repulsions between the negative charges of the peripheral phosphoryl groups and the jostlings caused by thermal agitation. At high enough temperature, the disruptive forces prevail and the two strands separate. By analogy with proteins, this phenomenon is called denaturation. An alternative term is melting. It is readily detected, for instance, by a sudden decrease in viscosity of the solution, explainable by the decrease in the rigidity of the threads. The temperature at which DNA denatures is called the melting point. It increases with the GC/AT ratio, because the GC pair is stronger than the AT pair. It is of the order of 85°C for human DNA.

Denaturation of DNA is a reversible phenomenon. If a solution of heated, denatured DNA is cooled somewhat below its melting point and kept at this temperature long enough to give complementary strands an opportunity to find each other, the molecules will rejoin. Understandably, the time required for this depends on the length and variety of the different molecules present. But even highly complex mixtures that contain thousands of different genes can be reconstituted under appropriate conditions.

We cannot watch this phenomenon in the cell, because no cell can stand a temperature at which its DNA starts denaturing. But just imagine yourself inside one of those cooling DNA solutions, observing the rebirth of beautifully undulating, semirigid, double-helical threads from the jumble of billions of intertwisted single strands. It is a mind-reeling spectacle. Innumerable temporary connections form by partial base-pairing, only to be torn apart again immediately by the violence of thermal agitation until chance brings together two complementary sequences that belong to authentic molecular partners. With lightning rapidity, the two strands spiral around each other, cementing their association before another turbulent collision comes to break it again. And so, one after the other, the double strands reform.

Were it not an experimentally established fact, we would never believe that this kind of molecular reunion by blind, random groping could ever take place with threads of such length and diversity. But it does, in a matter of hours, showing that, however hard we may try to penetrate it, the molecular world must necessarily remain entirely beyond the powers of our imagination owing to the incredible speed with which things happen in it.

In practice, reversal of denaturation has produced a valuable application known as hybridization. In principle, this method is designed to allow two different DNAs or, alternatively, a DNA and an RNA, to renature together. To the extent that long enough complementary sequences exist in the two species, they will join to form hybrid duplexes containing one strand of each. Assuming that such hybrid molecules can be recognized and evaluated—and several techniques allow this—hybridization can be used for estimating the degree of kinship between two DNAs or for fishing out a gene with the help of its transcribed mRNA, and so on. For example, an early experiment of this sort established that human DNA is very close to monkey DNA, less so to mouse DNA, even less so to fish DNA, differing more and more as evolutionary distance increases. It was one of the first applications of molecular biology to the study of phylogeny. In association with electron microscopy, hybridization has been turned into a tool of almost uncanny incisiveness and simplicity. Imagine two chains, A and B, complementary over their whole length except for a one-kilobase segment that exists in A—say, between bases 3,000 and 4,000—and has been excised from B. In the electron microscope, the A-B hybrid will appear as a double-stranded

thread bearing a single-stranded loop 0.3 μm long appended to it 1 μm away from one of its ends. Thus we can, with a simple ruler, find out how much has been removed from a chain and where (see pp. 308–309: mRNA splicing).

Transcription in the Nucleolus

From where we are, we can discern little more than the fine structure of the tangle in which we are caught. As luck would have it, when the nuclear envelope reformed around us, we found ourselves in one of the densest parts of the nucleus, a zone of what is called heterochromatin. The DNA in it is all bundled up and is functionally inactive. If we want to see some action, we must try to crawl out of this thicket and find one of the clearings marked on the nuclear maps left to us by those explorers who have looked at thin sections. They have described two kinds of openings: euchromatin (Greek *eu*, good), irregular areas where chromatin is dispersed instead of condensed as it is in the "other chromatin" and nucleoli, large, intranuclear organelles rich in protein and RNA. It will be helpful to move first to a nucleolus.

The term nucleolus is the Latin diminutive of nucleus, which, you remember, is itself the diminutive of *nux*. Thus, nucleolus literally means "small small nut"; it is a nucleus within a nucleus. It may have looked small to the early observers but, at our newly discovered molecular scale, the nucleolus appears quite large. It is a roughly spherical structure, irregularly outlined, about 2 μm in diameter—almost 7 feet at our millionfold magnification; we can just fit in. It has been known for more than a hundred years, and all sorts of structural details in it have been described. Observations were also made that suggested a major role of the nucleolus in growth and biosynthesis. The most impressive example is that of the egg cell of certain amphibians, in which the nucleus measures as much as 0.5 mm (0.02 inch), and as many as one thousand nucleoli may be seen in a single nucleus.

But nothing that classical observers were able to distin-

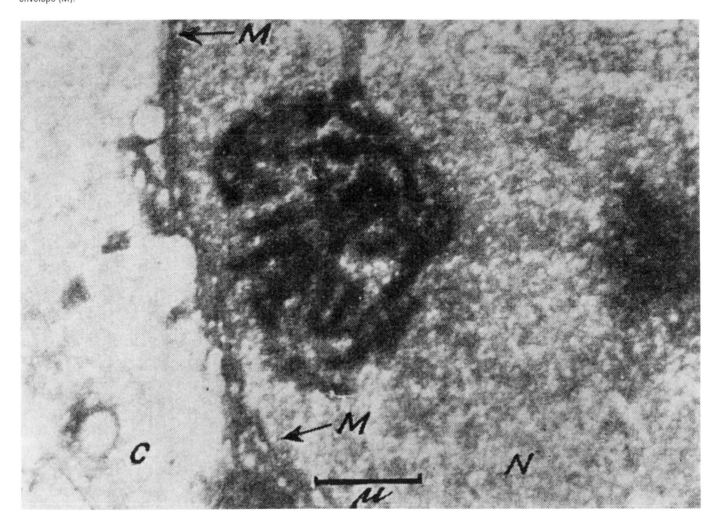

guish could have prepared us for the spectacle revealed by our molecular magnifying glasses. At first sight, it seems as if we have landed in the middle of some strange rain forest, overgrown with huge ferns garlanded by lianas. Or it could be the feather-lined nest of a giant bird. The whole space is crammed with long, delicately sculptured structures made of a central stalk, some 6 to 8 μm long (20–30 feet at our magnification) covered over a substantial part of its length with slender lateral digitations, or whiskers. Planted from 0.04 to 0.05 μm apart, these whiskers are of regularly increasing length, thereby giving the structure its fern-frond appearance. Each whisker is connected to the stalk by a knoblike bulge.

As many as a hundred or more such "fern fronds" may be found compressed within a single nucleolus. They are all strung together head-to-tail and joined by their stalks, which make up a single, uninterrupted thread, whose ends disappear in the dense heterochromatin brush that

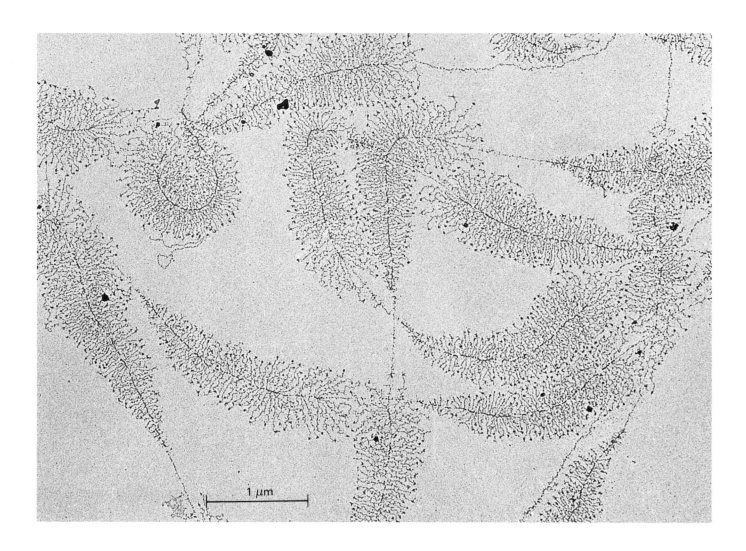

1 μm

borders the nucleolus. Bewhiskered and naked segments alternate regularly along this garland. To fully enjoy the beauty of this remarkable structure, we should really open the nucleolus, as some explorers have done, and allow the garland to uncoil freely. As it presents itself to us in the intact organelle, it forms an inextricable tangle. Count about half a mile for the main stalk, plus more than 10 miles total whisker length. It needs a lot of coiling to pack all this within a 7-foot box. The nucleolus may look like a clearing to somebody crawling out of the surrounding heterochromatin. But it is a far cry from what one might call an opening. Even so, it provides us with our best opportunity for witnessing a phenomenon of absolutely central importance, and we must take full advantage of it. Look closely, and you will see that each fern frond is throbbing with activity. The knobs move, the whiskers lengthen, the stalks rotate, all in perfect synchrony.

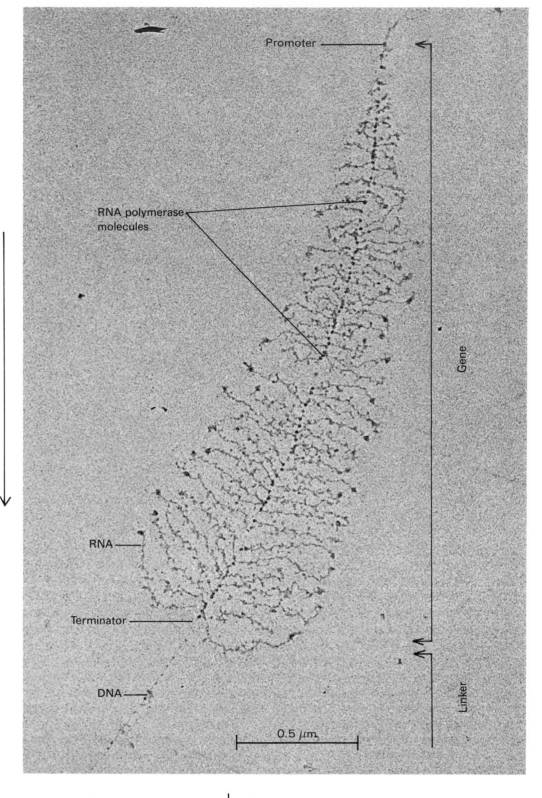

Direction of transcription

Promoter

RNA polymerase
molecules

Gene

RNA

Terminator

DNA

0.5 μm

Linker

Functional significance of the "fern frond." The knobs on the central stalk are molecules of RNA polymerase in the process of transcribing a gene. The stalk is DNA; the progressively lengthening whiskers are growing RNA transcripts. The bare part of the stalk is an untranscribed linker between genes.

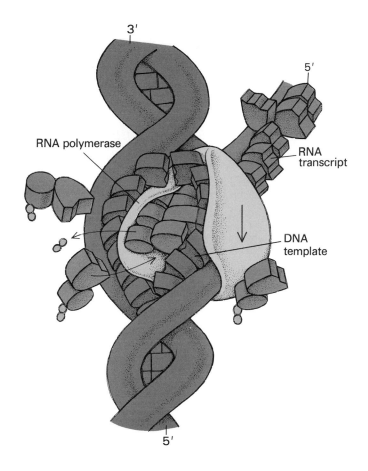

3'

5'

RNA polymerase

RNA
transcript

DNA
template

5'

A view of the transcription process. The regular double-helical structure of DNA is disrupted locally by the insertion of RNA polymerase. This enzyme wraps around one of the DNA strands and uses it as template for the assembly of a complementary RNA chain from nucleoside-triphosphate precursors. The enzyme moves along the template DNA in the 3' to 5' direction. RNA lengthens in the 5' to 3' direction (tail growth). DNA resumes its double-helical configuration as the enzyme moves on.

We can gain some clues to the meaning of these concerted motions by watching the knobs. They course along the central stalk in the direction of increasing whisker length at the appreciable speed of about 1 μm per minute, which, magnified a millionfold, amounts to two-thirds of an inch per second. As the knobs move, the whiskers attached to them increase in length in a manner that is exactly commensurate with the distance traveled along the stalk. When a knob reaches the base of the fern frond, it falls off and releases its attached whisker, which by then is from 4 to 5 μm long. At the same time, a fresh knob binds to the stalk at the tip of the frond and starts sprouting a new whisker. What we are watching, therefore, is some sort of *perpetuum mobile* of knobs along the stalk of each fern structure and, associated with it, the growth and release of whiskers. Each fern frond turns out some 20 to 25 whiskers per minute, to give a total for the whole nucleolus of 2,000 or more per minute. In length, this adds up to about 1 cm (5–6 miles at our magnification). This material fills the spaces between the bundled-up fern fronds, where it undergoes a complex form of processing before being conveyed to the cytoplasm.

With better adjustment of our focus, we have no difficulty recognizing the main components of this remarkable factory. The central stalk, semirigid and 2 nm thick, shows the characteristic double-helical structure of DNA; it is a fully extended DNA thread. The lateral whiskers are thinner and more flexible; they are made of single-stranded RNA. The knob connecting the RNA whisker to the DNA stalk is a voluminous protein, wrapped collarwise around the DNA. This protein projects an inner spur between the two DNA strands and keeps the 3' end of the RNA whisker closely aligned against one of the strands. The process we are looking at is none other than the biosynthesis of RNA. The knob at the growing end of the chain is the enzyme that catalyzes this process. Its association with DNA reflects the fact that it is guided in the performance of its biosynthetic job by instructions supplied by the DNA. We have, in fact, just been afforded our first glimpse of transcription.

Chemically, the growth of an RNA chain is a simple process. It takes place by successive nucleotidyl transfer

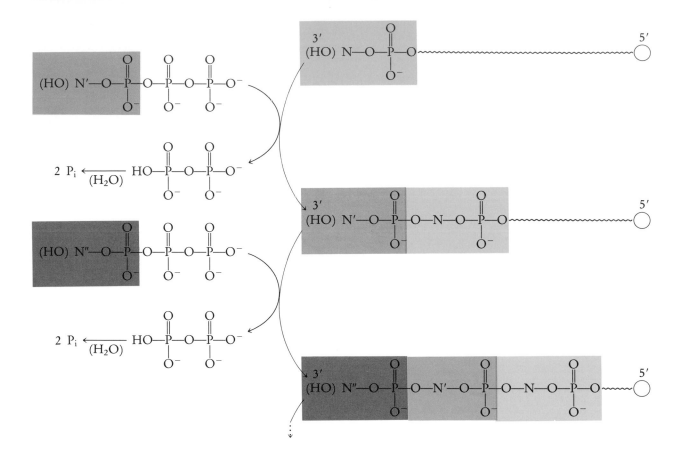

from a nucleoside triphosphate to the 3'-terminal hydroxyl group of the growing chain, as indicated in the scheme shown above. It is an instance of a single-step biosynthetic process dependent on a β_p attack (Chapter 8). Note that we are dealing here with a case of tail growth: activated building blocks are added one by one to the growing chain. This is in contrast to protein synthesis, which takes place by head growth—that is, by repeated transfer of the activated growing chain to (activated) building blocks.

The enzyme that synthesizes RNA is called RNA polymerase. Its substrates are ATP, GTP, CTP, and UTP, which may be considered the corresponding nucleoside monophosphate building blocks AMP, GMP, CMP, and UMP, activated by two successive γ_d phosphoryl transfers from ATP:

$$NMP + ATP \longrightarrow NDP + ADP$$
$$NDP + ATP \longrightarrow NTP + ADP$$

The biosynthetic process releases pyrophosphate, which, as is customary, is hydrolyzed to two P_i. In other words, to add an NMP to the 3' end of a growing RNA chain, the cell uses two terminal pyrophospate bonds of ATP, or 28 kcal per gram-molecule.

So far so good. But there remains the problem of which of the four available nucleotidyl groups to add to the 3' end of the growing chain at each step. Instructions to this effect come from the DNA by way of the magic cipher. The RNA polymerase is constructed in such a way as to align its two substrates, the growing chain and the donor NTP, in antiparallel fashion alongside a DNA strand. It will catalyze the transfer only if perfect complementarity obtains between the paired bases. As soon as the right NTP comes along and its nucleotidyl group is added to the growing chain, the enzyme moves one notch, to bring its active center back in the right strategic position with respect to the new 3'-hydroxyl end of the growing RNA chain.

In this process, the DNA instructions are read, base by base, in the 3'⟶5' direction, while the RNA strand grows in the 5'⟶3' direction. Transcription proceeds until the enzyme encounters a terminator on the DNA— a specific base sequence that somehow causes the enzyme to stop. A special factor, known as rho, helps the enzyme off the DNA and releases the RNA, with the concomitant hydrolysis of ATP, which presumably provides the necessary energy. To start a new chain, the enzyme needs the assistance of another factor, called σ, which guides it to a special initiating binding site on the DNA, signaled by a particular base sequence called promoter. Unlike DNA polymerase, which will be considered in the next chapter, RNA polymerase has no need for a preformed chain to serve as first acceptor. It can use a single nucleoside triphosphate: in most cases, ATP or GTP.

If we put all this information together, we may now identify the DNA segment that is being transcribed by RNA polymerase as a gene—a chemically coded element of genetic information. Transcription is the mechanism whereby this information is copied in RNA language. It is the obligatory first step in the expression of the gene into what, in classical genetics, is called a phenotypic character (Greek *phainein*, to appear). Our nucleolar gene has a length of from 4 to 5 μm, or from 12,000 to 15,000 base pairs. It is flanked on its 3' end by the promoter sequence and on its 5' end by the terminator sequence. Beyond the terminator sequence is a stretch of untranscribed DNA—the naked part of the stalk—up to several thousand base pairs long, which ends with the promotor sequence of the next gene. This segment is called a spacer or linker. It may play a variety of roles in gene regualtion in addition to bearing the promoter and terminator sites.

Nucleolar transcription is a unidirectional, one-sided process: only one of the two DNA strands is transcribed. This, we will see, is a general, though not absolute, property of transcription. Its rate of progress is of the order of 1 μm per minute, which means that the RNA chain grows at the rate of fifty nucleotides per second, or that it takes RNA polymerase no more than two-hundredths of a second to try various nucleoside triphosphates for fit, find the right one, substitute one bond for another, move on

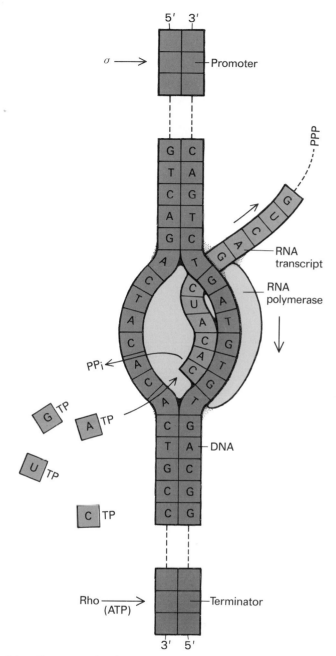

Schematic representation of transcription. RNA polymerase separates the two strands of DNA and assembles an RNA chain complementary to one of the strands (and therefore identical in sequence with the other strand except for the replacement of T by U). The initial insertion of the enzyme is determined by a special promoter sequence and is helped by the sigma factor. A terminator sequence signals the detachment of the enzyme, which is assisted by the rho factor and requires ATP. Note that the 5' end of the RNA strand bears a nucleoside triphosphate that has served as the initial acceptor.

through the next base pair in the DNA double helix, and get ready for another round of the same kind. Once again, the speed of molecular events leaves us baffled. But wait until the next chapter. DNA replication is much faster than its transcription, at least in prokaryotes.

Such speed is all the more astonishing in that the polymerase has to follow a helical path around the DNA thread. You can readily appreciate that, with ten nucleotide pairs per turn of the DNA double helix, the enzyme must revolve around the DNA three hundred times per minute as it moves on. It is hard to imagine the whole bulky set of knobs, with their trailing hairs of growing RNA, whirling around the DNA at that speed. Except, of course, that we must beware of our imagination when it comes to the molecular world. How many times has that lesson been thrust upon us already during our visit. But there is an alternative. Instead of the polymerase circling around the DNA, the DNA could rotate inside the polymerase. Here, however, another topological problem is encountered: the DNA is not free to rotate. It is part of a very long loop belonging to a chromosome and, in addition, it is all twisted upon itself to fit within the exiguous space offered by the nucleolus. Rotation of such a coiled-up thread can take place only in separate short segments provided with appropriate swivels. Several of the bonds in a polynucleotide chain allow free rotation and can serve as swivels. But this will work only with a single strand, not with a double strand. The cell's way of getting out of this quandary is simple: nick one of the strands, and the remaining, intact strand can serve as swivel for the whole double-stranded structure. As long as the nick is on only one strand, the molecule runs no risk of breaking apart; the loose strand is kept coiled around the intact one by the various forces that stabilize the double helix.

In practice, if we were able to watch transcription in slow-motion replay, we would see the RNA polymerase molecules cutting through between the two strands of the DNA, causing "positive supertwisting" of the double helix in front, and "negative supertwisting" (undertwisting) behind. These two twists cancel each other between any two polymerase molecules and are, therefore, manifest only at the two ends of the gene. A nick on one of the

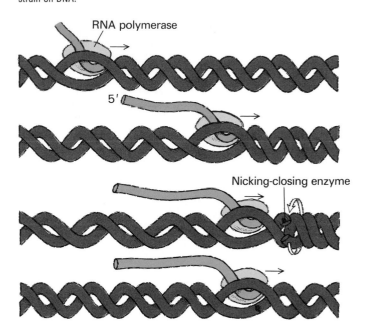

Topology of transcription. The progress of RNA polymerase causes the positive supertwisting of DNA in front and its negative supertwisting behind. A nicking-closing enzyme relieves the strain on DNA.

RNA polymerase

5'

Nicking-closing enzyme

strands near the promoter and terminator regions suffices to relax the resulting constraints, except for what is needed to provide the DNA with its rotation energy. RNA polymerase is remarkably economical in this respect. It supports the whole of its mechanical, as well as its chemical, work with the 28 kcal it has available at each step. It seems also to be able to carry out initiation, which requires slipping through about six base pairs, without any extra supply of energy. Extricating it again does, however, cost energy. Rho, as we have seen, consumes ATP in the performance of its job. Another remarkable property of RNA polymerase is its low record of mistakes, which has been estimated at about one wrongly inserted nucleotide in ten thousand. Apparently, cells can afford this level of "noise" in transcription, which, unlike duplication, does not seem to be subject to proofreading.

Now that we have some idea of how nucleolar transcription works, let us take a look at its product. It is a long RNA chain of about fourteen thousand nucleotides, 4.5 million daltons in molecular mass, 45S in Svedberg units. The first striking thing about it is that all molecules are identical. This means that the hundred or more genes that are strung along the nucleolar DNA specify the same message. What we are encountering in the nucleolus is a typical example of genetic amplification, or multiple-copy genes. It is a device used by the cell when the demand exceeds the transcription capacity of a single gene. Assume that the cell needs a thousand molecules of this particular kind of RNA per minute, which is the order of magnitude for our "average" mammalian cell. If it takes 5 minutes to make an RNA molecule, five thousand RNA polymerases must be put to work. They cannot possibly all crowd onto a single gene. The only solution, therefore, is to multiply the number of genes. In certain animal species, this process is itself genetically controlled. At certain stages of extremely heavy demand, the cells rush their nucleolar RNA production by selectively turning out numerous extrachromosomal copies of the genes involved.

If we now examine the structure of the RNA made in the nucleolus, we will have no difficulty recognizing some familiar sequences. We have encountered them before in ribosomes. In fact, the nucleolar RNA contains the whole of the ribosomal RNA except for the 5S component, but in the form of a single large chain. This chain must be processed to yield the 18S piece of the small ribosomal subunit (2,000 nucleotides) and the 28S (5,000 nucleotides) and 5.8S (160 nucleotides) pieces of the large ribosomal subunit. Like proteins, RNA molecules have a history, which is programmed in their sequence.

We can get some idea of how this programming operates by watching a growing RNA chain while it is still being elongated. It is, you will notice, assailed by a swarm of enzyme molecules that zero in on certain specific sites, no doubt attracted there by some signal. These are transmethylases. They transfer methyl groups from the activated methyl donor, S-adenosylmethionine (Chapter 8), onto certain ribose molecules in the growing RNA chain. As it happens, methylation affects only those segments of the RNA chain that are destined to participate in the construction of ribosomes. After the 45S RNA is completed and released, the nonmethylated parts are bro-

Processing of ribosomal RNA (rRNA) in the nucleolus. While transcription of the corresponding gene (rDNA) proceeds, portions of the newly made RNA transcript destined to form rRNA are progressively methylated. After detachment of the completed transcript as a 45S precursor, ribonuclease action cuts off the unmethylated segments, leaving mature rRNAs. These mature molecules then combine with proteins and are delivered into the cytoplasm, where they are joined by additional proteins and by a small 5S rRNA not made in the nucleolus, to form the two ribosomal subunits.

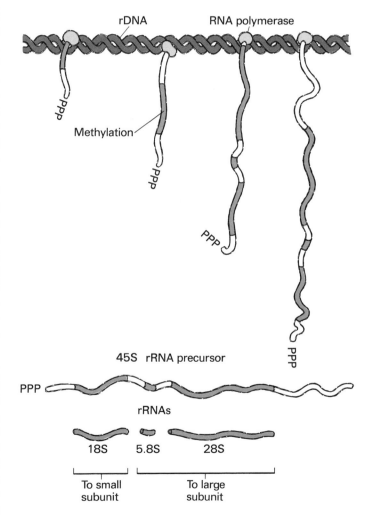

ken down by ribonucleases. But the methylated segments remain intact, protected by their methyl groups. Thus, the signal that attracts the transmethylases provides a mechanism for guiding the process whereby the various rRNAs are cut to size and completed. It is likely that this signal lies in certain base sequences that serve as binding sites for the transmethylases. These base sequences are themselves transcribed from the corresponding DNA, in which, therefore, the future of the molecule is genetically encoded.

But time once again is pressing us. We must move on. Before we do so, however, let us cast a last look around. It is remarkable how our perception has changed. When we first arrived, we were impressed with shapes and forms—fern fronds, feathers, knobs, stalks, whiskers—to the point of having a strain put on our morphological vocabulary. Early explorers had the same problem. They used other words, such as nucleoloneme, nucleolar organizer, fibrillar zone, granular zone, but their purpose was the same: descriptive. And so was the effect: a mixture of wonder and puzzlement.

How different our present molecular view. The nucleolus is simply a large ribosomal RNA factory, built by tens of thousands of molecules of RNA polymerase, transmethylase, ribonuclease, and other enzymes, gathered around a long, unfolded loop of DNA that bears multiple copies of the same gene and supports thousands of distinct transcription units operating simultaneously. Every second, this factory assembles, and methylates in certain specific sites, some fifteen to twenty RNA strands about fourteen thousand nucleotides long. These are then cut and trimmed into three pieces totaling about half this length: the 18S, 28S, and 5.8S rRNAs. Newly made ribosomal proteins, synthesized in the cytoplasm and admitted into the nucleus by the pores, come and bind these RNA pieces and return them to the cytoplasm to participate in the assembly of an equivalent number of new ribosomes. Wear and tear of ribosomes is thereby compensated, and the protein-synthesizing potential of the cell is maintained unimpaired. In times of rapid growth (e.g., in the amphibian egg cell) this manufacturing potential may be stepped up many times by amplification of

genes and multiplication of nucleoli to the point of turning out as many as one million or more new ribosomes per minute. When a cell prepares for division, the nucleolar factory packs up its gear and folds back into its chromosomal shelter. In the completed daughter nuclei, it emerges again from the same chromosomal region, which corresponds to the nucleolar organizer of the early observers.

Transcription in Euchromatin

Visiting the nucleolus has greatly simplified the next part of our tour. As we worm our way through the loosened coils of euchromatin, we have no difficulty recognizing what is going on. Everywhere, our eyes fall on the same familiar sight of RNA polymerase molecules trailing nascent RNA threads along rapidly rotating DNA fibers. Euchromatin is simply a conglomerate of sites where nuclear DNA is being transcribed into RNA. In most nuclei, the organization of these transcription sites is almost impossible to unravel. But, as is often the case, Nature has favored us with an exemplary model in the so-called lampbrush chromosomes that are seen in many oocytes at the diplotene stage of the first meiotic prophase (see Chapter 19), where paired, duplicated chromatids, still joined by their chiasmata, are actively transcribed on exactly the same sites. These transcribed sites make fluffy lateral loops—perfectly symmetrical, owing to the close fit between the duplicated chromatids—that extend from the axial stem of the lampbrush. This stem is made of untranscribed DNA segments and of the protein skeletons of the chromatids. It is clear from this structure that each of the two chromatids consists of a single DNA thread, of which certain stretches are transcribed and others are not. This is, in fact, what we would see, but without mirror image, should we have the possibility of following a chromosomal DNA thread over some distance in the usual interphase nucleus. Lampbrush chromosomes also give us an interesting glimpse of the manner in which the DNA is anchored as loops to the protein framework. However, we distinguish only the loops that are unfolded

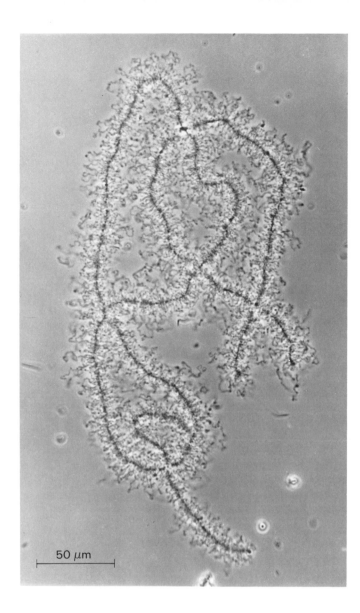

Phase-contrast micrograph of a lampbrush chromosome bivalent from an oocyte of the newt *Notophthalmus viridescens*. Hundreds of transcriptionally active loops project from the main axis of each chromosome. The two homologous chromosomes are joined by several chiasmata.

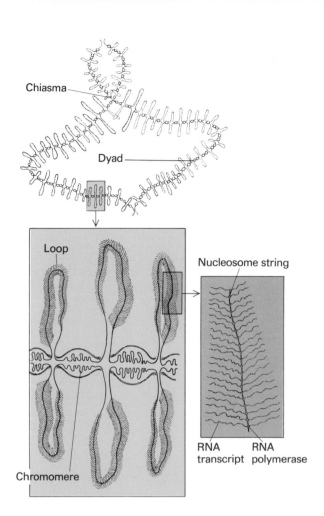

Diagrammatic representation of the structure and function of lampbrush chromosomes characteristically found in oocytes during the extended prophase of the first meiotic division (see Chapter 19). The drawing at the top shows the typical appearance that can be discerned in the light microscope, as shown at the left: fluffy, symmetrical loops alternate with thicker condensations called chromomeres, recalling the shape of a brush used to clean the chimneys of kerosene lamps. Two identical-looking chromosomes are joined together by one or more chiasmata, which are the sites of recombination during crossing-over (see the illustrations on p. 287). As shown in the yellow enlargement, each chromosome is really a dyad, made of two identical sister chromatids resulting from the duplication process that has preceded the onset of meiosis. Each chromatid consists of an uninterrupted thread supported by a protein framework. Portions of this thread are fully extended to form the loops; the others are bundled up into the chromomeres. Special treatment allows the loops to be visualized in the electron microscope (see the micrograph on p. 304, which shows part of a loop) as possessing the typical fern-frond appearance of an uncoiled string of nucleosomes in the process of being transcribed (blue enlargement). Note the change in direction of transcription indicated in the yellow enlargement. It illustrates the fact that either one or the other strand of the DNA duplex may be transcribed, but rarely both.

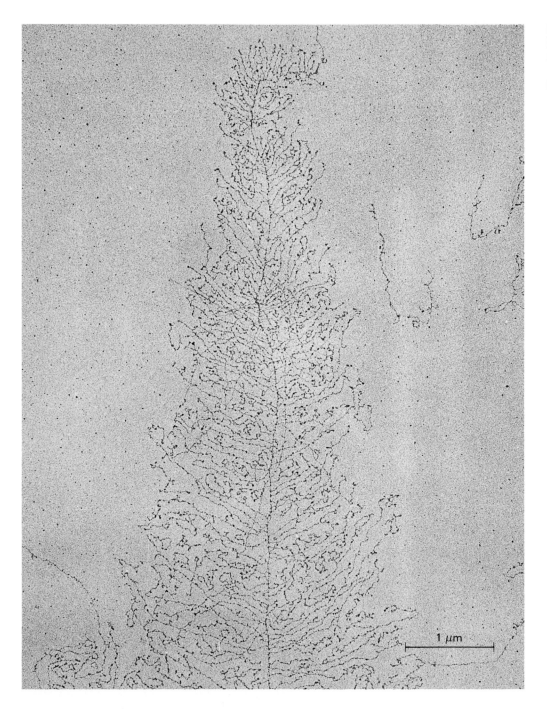

Electron micrograph of dispersed fibrillar material from lampbrush chromosome of *Triturus viridescens* oocyte. Compare with the micrographs on pages 295 and 296.

1 μm

and transcribed, not those that are compacted into the stem. We will see more details of chromosome organization in the subsequent chapters.

Part of euchromatin transcription complements nucleolar transcription in the upkeep of the protein-synthesizing machinery. Here, in a corner, a string of small, identical genes turns out 5S rRNA for the large ribosomal subunit. In another area, tRNA molecules are being made, readily recognized by their shortness, by their peculiar cloverleaf folding, and especially by the number of different enzymes clinging to them and producing the numerous chemical modifications so characteristic of this type of RNA. Also made in multiple copies are certain small RNA molecules (sRNA) that are beginning to emerge as cofactors in such important processes as protein signaling (Chapter 15) and mRNA splicing (see p. 310). Much of the RNA made in euchromatin, however, is mRNA. It is made by a special RNA polymerase programmed to obey a particular set of signals, which includes a TA doublet, known as the TATA box, some thirty base pairs upstream from the starting point, as part of its promoter sequence.

By definition, messenger RNAs are tapes in which the amino-acid sequences of given polypeptide chains are written in code and are complemented by various signals that ensure proper positioning and reading by the protein-synthesizing apparatus. We have already noticed that mRNA molecules flow out continuously through the nuclear pores and therefore must be made in the nucleus. We are now finding out that they are made by transcription of nuclear genes. Ultimately, therefore, the structures of the cell's proteins are inscribed in DNA sequences. And this is how the genes (apart from the small number that specify noncoding, functional RNAs such as rRNAs, tRNAs, and sRNAs) control hereditary characters. They do so by way of their translation products, the proteins, which themselves, as we have seen, mediate most of the manifestations of life.

As in nucleolar transcription, only one of the two DNA strands is usually transcribed into mRNA. This is understandable. If both strands were transcribed, the product would consist of two complementary RNAs, which, at the first opportunity, would lock each other into a double helix prevented from participating in further information transfer. Even if this mutual block could be overcome, there is little chance that such long, complementary RNA strands could both be "meaningful"; that is, would each translate into a functional protein molecule. Two-sided transcription, therefore, would not be just highly cumbersome; most of the time it would also be useless. This does not mean that the same strand of DNA is always being transcribed. Transcription templates occur sometimes on one, sometimes on the other of the two strands of a DNA thread, and the direction of transcription therefore changes from one part of a chromosome to another. Note also that the ban against two-sided transcription is not absolute. Especially in bacteria and in viruses, several examples are known of genes facing each other on the same piece of DNA. Our own mitochondria, in a remarkable feat of economy possibly inherited from their putative prokaryotic ancestors, have succeeded in cramming transcribable information on both strands of their DNA and even in eliminating most of the spacers.

An Unexpected Editorial Mechanism

In bacteria, which do not possess a fenced-off nucleus, but have only a single, circular chromosome in direct contact with the cytoplasm, startling evidence of the relationship between DNA, RNA, and protein can be seen in the form of what may be called "composite fern fronds," structures in which the RNA whiskers produced from the DNA stalk by RNA polymerase are themselves read off by protein-synthesizing ribosomes. This can happen because both the growth and the reading of mRNA are in the 5′——→3′ direction. Thus, it is possible for reading to start before the chain is completed. Such structures illustrate in striking fashion the principle of co-linearity between the DNA gene, or cistron, its transcription product (mRNA), and its primary translation product (unprocessed polypeptide).

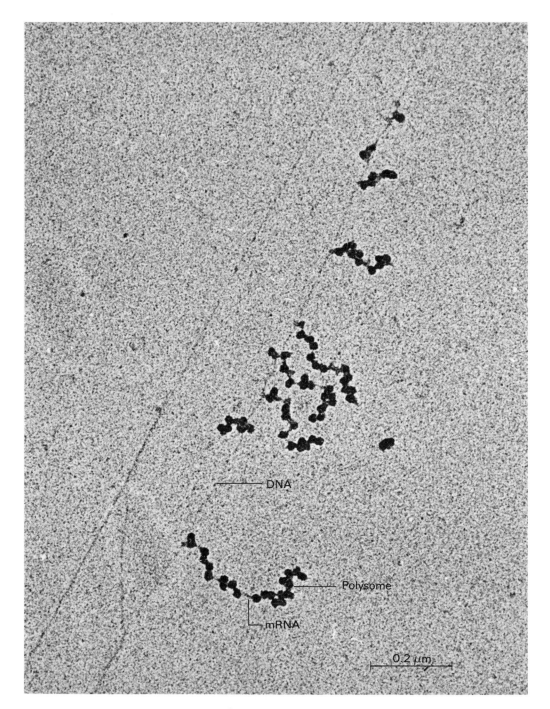

Coupled transcription-translation in bacteria. This electron micrograph of dispersed fibrillar material from *E. coli* shows ribosomes strung on RNA strands that grow from DNA in the process of being transcribed.

DNA

Polysome

mRNA

0.2 μm

Schematic interpretation of the electron micrograph on the facing page. The growing polypeptide chains are beyond the resolving power of the technique used and are not visualized in the electron micrograph.

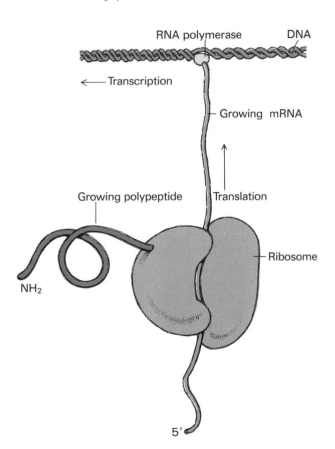

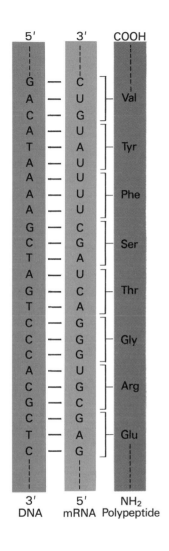

Similar figures cannot be seen in eukaryotic cells, where transcription and translation are separated by the nuclear envelope. But it never occurred to anybody to doubt for a single moment that co-linearity applied equally well to eukaryotes, so perfectly logical did it appear to be. To most people acquainted with the mechanisms of transcription and of translation, co-linearity was a self-evi-

dent necessity. A scrambled message was unthinkable, for where would the unscrambler be?

But once again—molecular biology is fertile in such surprises—the unthinkable has proved true. If you watch what happens to most primary transcription products made in euchromatin, you will see that they undergo a considerable degree of surgery before being delivered into the cytoplasm. In a process reminiscent of cutting a movie film or editing a tape, a number of segments are excised from the RNA ribbon and the remaining ones are stitched back together. The pieces that are kept in the mature mRNA are called exons because they are <u>ex</u>pressed; those that are removed are named introns, for <u>inter</u>mediate or <u>inter</u>vening sequences. The extent to which genes are split into fragments in this way may be considerable. For instance, a gene in the hen oviduct that codes for conalbumin, one of the proteins of egg white, consists of seventeen exons separated by sixteen introns; as many as fifty exons have been counted in a collagen gene. Introns are readily seen when the hybrid duplex between the mRNA and its gene is examined in the electron microscope. They appear as single-stranded (DNA) loops of various sizes appended to the double-stranded heteroduplex.

As indicated by the shape of such duplexes, cutting and splicing mRNA is a little simpler than cutting a movie in that the exons are united in the order in which they follow each other in the original gene. At least the message is not entirely scrambled, and co-linearity is respected to some extent. You can easily see how this is done by looking at the primary transcription product (first described as heterogeneous nuclear RNA, or hnRNA, and now designated mRNA precursor, or pre-mRNA) as it is being processed. You will notice that the introns form loops that are closed in such a way as to bring together the 5' and 3' ends of two adjacent exons. Excision of the intron and suturing of the two exons then take place in immediate succession. There is no opportunity for the severed exons to separate and become scrambled.

There is, however, something strange about the manner in which these loops are closed. They are not joined by complementary sequences, as are the usual RNA loops, for the RNA parts involved are not complementary, as they should be if they served to close the loop (p. 257). Therefore, these parts must be recognized by "something else" acting as a pincer or vise while the splitting and splicing enzymes do their jobs. Indeed, the terminal sequences of many introns are, if not identical, at least closely similar, sufficiently to be recognized by the same structure. Unfortunately, we cannot distinguish

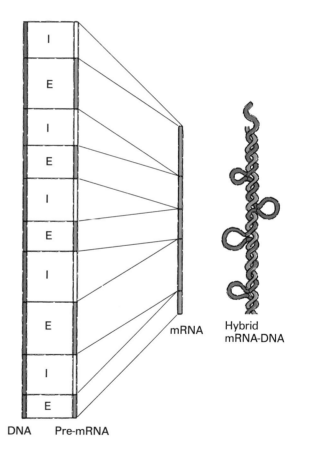

Schematic representation of mRNA processing. Exons (E) are segments that are expressed. Introns (I) are intermediary segments that are excised. Note that exons are spliced together co-linearly, as shown by the mRNA-DNA hybrid. DNA introns appear as loops on the duplex between mRNA and DNA exons.

DNA   Pre-mRNA

mRNA   Hybrid mRNA-DNA

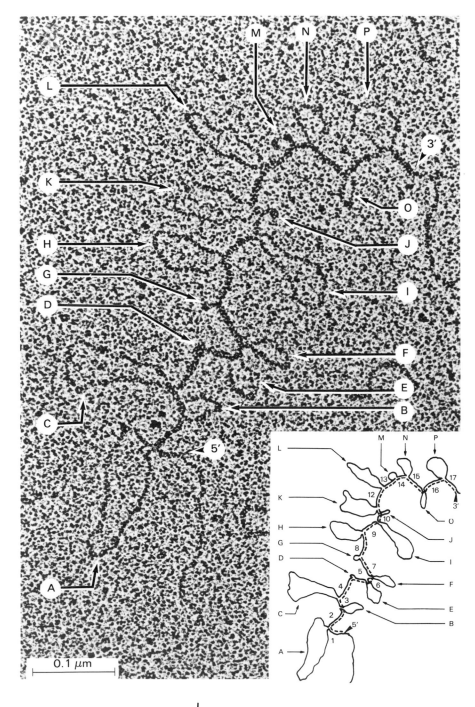

Electron micrograph of mRNA-DNA hybrid between conalbumin (from egg white) gene and mature mRNA. The thick lines, identified by numbers, are hybrid regions (exons). The thin loops, identified by letters, are DNA segments corresponding to introns that have been excised from pre-mRNA upon splicing. (Compare with the diagram on the facing page.)

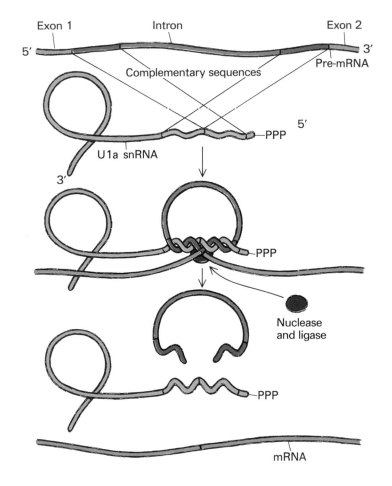

Exon 1   Intron   Exon 2

5'   Pre-mRNA   3'

Complementary sequences

5'

PPP

U1a snRNA

3'

PPP

Nuclease and ligase

PPP

mRNA

ent proteins seem to be involved in these complex operations, in addition to the putative snRNA vise. They form little bundles that can be seen strung all along the pre-mRNA threads. They can also be isolated as 30S ribonucleoprotein particles (RNPs) after selective cutting of the threads.

Some other RNAs in addition to nuclear pre-mRNAs undergo splicing, apparently by different mechanisms. Most amazing is a recently discovered pre-rRNA that seems to do its own editing, including the shedding of an intron and the subsequent splicing, without any outside help. Some mitochondrial mRNAs, as we shall see, rely on their introns to generate the necessary help.

Besides splicing, mRNA molecules must also be fitted with their characteristic 7-methyl-GTP cap, additional methyl groups, and poly-A tail. The latter addition is usually preceded by substantial exonucleolytic trimming of the 3' end. These reactions likewise take place in the nucleus, probably before splicing. Capping may even start before transcription is terminated. The finished mRNAs are then delivered into the cytoplasm through the nuclear pores in combination with special carrier proteins.

So far, split transcripts have not been encountered in prokaryotes. On the other hand, not all eukaryotic transcripts are split. For instance, the histone mRNAs represent primary transcription products, as do bacterial mRNAs. They also resemble the latter in lacking a 7-methyl-GTP cap and poly-A tail. Surprisingly, at least one mitochondrial gene (in yeast) has been found to be split.

The discovery of gene splicing has given molecular biology one of its rudest jolts so far. When light first fell on the genetic organization of primitive forms of life in the late 1950s and early 1960s, the picture glimpsed was one of satisfying orderliness. Explorers were led to think of genes as forming a neatly catalogued library equipped with retrieval mechanisms that could, in spite of their complex network of interacting feedback loops, be accommodated within an intuitively simple conceptual framework. Today, with more books read and, especially, with exploration extended to the eukaryotes, this

clearly what does the joining. It could be a special kind of RNA, known as small nuclear RNA (snRNA), a group of short-chain molecules from 90 to 220 nucleotides long, endowed with several characteristic structural features. One of these snRNAs (U1a) possesses a continuous sequence of bases, of which one part is near-complementary to the 3' end, the other to the 5' end, of introns. It could serve to align these two ends against each other in such a way as to position the exon-intron connections accurately for splitting by a ribonuclease, followed by splicing of the exons by a ligase (an enzyme similar to the DNA ligase that we will encounter in Chapter 17). Some thirty differ-

Relationship between products of spliced and unspliced messages when pre-mRNA includes a facultative intron.

A. The intron is fully translatable. The two proteins have the same N-terminal sequence. They also have the same C-terminal sequence if n, the number of nucleotides in the intron, is a multiple of three (1). If not, the C-terminal sequences are different (2).

B. The intron includes a termination codon. After splicing, reading continues until the next termination codon. The two proteins have identical N-terminal segments. The T and t proteins of SV40 illustrate this case (see the illustration on p. 312).

C. The intron includes an initiation codon. Its removal promotes an AUG (or GUG) codon situated downstream to function as initiation codon. The resulting protein and the C-terminal part of the product of translation of the unspliced mRNA have identical sequences if m, the number of nucleotides separating the two initiation codons, is a multiple of three—that is, if the reading frame is unchanged (1). If not, the two sequences are entirely different (2). The VP proteins of SV40 illustrate these two possibilities.

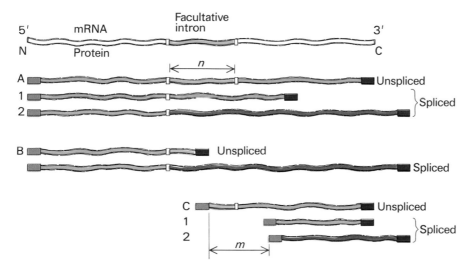

vision has been shattered. No longer a librarian's dream, it could even become a cryptographer's nightmare. It now appears that most eukaryotic genetic messages are written in pieces separated by stretches of untranslated DNA. If we have no splicing key, the messages are completely ununderstandable, as many an explorer has found after searching the nucleus directly for information. But cells can hardly derive much evolutionary advantage simply from keeping snoopers out. Surely there must be more substantial benefits to compensate for such a cumbersome, hazardous, and costly way of storing information. Otherwise, this mechanism could hardly have resisted the selective pressures that must have worked against it.

One major advantage of splitting genes, it has been pointed out, is that it enormously enhances the scope of evolutionary experimentation by recombination (see Chapter 18). Exons can be reshuffled at a high rate by means of their appended introns, and it matters little where or how exactly the joining is done. As long as the critical sequences that govern splicing are not altered, the

final edited product will consist of neatly connected exons. What makes such a mechanism particularly valuable is that exons do not seem to be just randomly cut segments of the genetic message but may well correspond to a defined structural or functional domain of a protein molecule, to a "miniprotein," as it has been called. The odds of coming up with something interesting by reshuffling such pieces, even in a haphazard fashion, are appreciable, just as is the probability of generating a meaningful sentence by restringing complete words or phrases.

Another, more immediate, advantage of gene splicing derives from the possible modulation of the splicing mechanism itself by means of what might be called facultative introns (intervening sequences that may be either spliced out of a pre-mRNA or left in it). If such an intron contains the initial part of a message, its removal will promote another AUG or GUG codon, situated downstream, to the rank of initiating codon. Depending on whether this new start signal belongs to the same reading

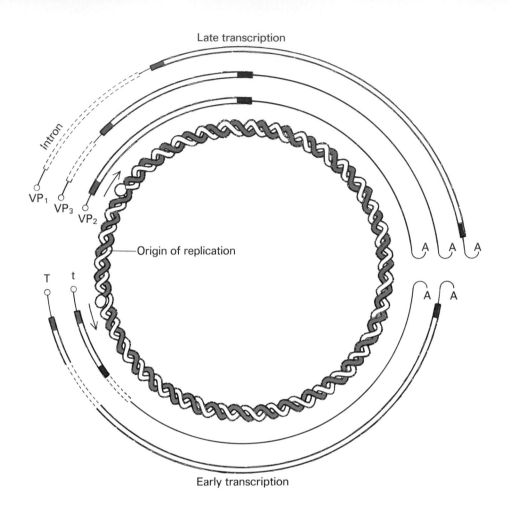

Late transcription

Intron

VP₁ VP₃ VP₂

Origin of replication

T   t

Early transcription

The small virus SV40 (see Chapter 18) consists of a circular double-stranded DNA, 5,224 nucleotide pairs long. Its transcription takes place in two stages. Thanks to distinct modes of splicing, the two resulting pre-mRNAs give a total of five translation products. First about one-half of the genome is transcribed from one strand. The other half is transcribed later from the other strand. The early pre-mRNA is spliced in two different ways to give two translation products, t and T, which have the same N-terminal sequence of 100 residues. T is much longer than t because the stop codon that terminates translation of the t mRNA has been spliced out of the T mRNA, allowing translation of the message up to the next stop codon, situated almost at the extremity of the molecule. The late pre-mRNA is also spliced in two different ways; in addition, it is translated in unspliced form. In VP₃ mRNA, the initiation codon of VP₂ is removed with an intron: translation starts further down, actuated by an initiation codon that happens to be in phase with that of VP₂. Therefore, the VP₃ protein is identical with the C-terminal part of the VP₂ protein. In VP₁, the two initiation codons are spliced out; translation is initiated by a codon that is out of phase with the others, so that the VP₁ protein has nothing in common with the other two VP proteins.

frame as the excised one, the resulting protein will be identical with the C-terminal part of the product of translation of the unamputated message or it will be completely different. If, on the other hand, the facultative intron includes a stop signal, its removal will allow read-through until the next stop sign, with synthesis of a protein that now shares its N-terminal sequence with the product of the intact message. Finally, if the facultative intron is fully translatable, in register with the initial part of the message, the two proteins will also have the same N-terminal sequence and they may possess a common C-terminal sequence. This will happen when the reading frame of the subsequent exon is not altered by removal of the intron—that is, when the length of the intron, measured in number of nucleotides, is an exact multiple of three.

It is remarkable that, with only a few messages decoded so far, examples of all the possibilities mentioned have already been found. A small DNA virus known as SV40 (see Chapter 18) generates as many as five different proteins from only two primary transcripts. Immunoglobulins of the IgM class are made in two forms, one soluble, the other membrane-bound thanks to an additional hydrophobic C-terminal sequence that results from the removal of an intron containing a stop codon. A particularly intriguing example of alternative splicing of the same pre-mRNA has been observed in yeast mitochondria. These organelles apparently manufacture self-adulterating messages whose primary translation products (maturases) induce further splicing of the mRNA to a form coding for a piece of the respiratory chain.

Even in the prokaryotic world, things are not as simple as they were thought to be. Overlapping genes seem to be quite common, and at least one case is known (in a small bacterial virus) of three genes overlapping in such a way that the region of overlap in the DNA is actually translated in all of its three distinct reading frames.

So far, transcript splicing has not been observed in bacteria, whereas it seems to take place in all eukaryotes, even the most primitive. Apparently, therefore, the development of this process coincided with the great prokaryote-eukaryote transition and may even have played a key role in it. But we must beware of the obvious. It could be argued—and has been quite cogently—that genes are more likely to have started in the form of small pieces that were continuously rearranged in a more-or-less haphazard manner than as the highly compact, often polycistronic, sequences characteristic of present-day prokaryotes. If such is the case, then eukaryotes could be viewed as having tamed this ancient process and turned it into a mechanism of great precision while conserving its versatility, and the prokaryotes as having squeezed it out in the course of evolution, thereby gaining generation rapidity but losing the genetic flexibility that has allowed eukaryote development.

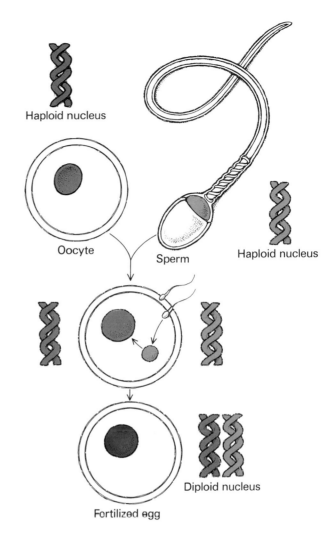

Schematic representation of fertilization illustrates the formation of a diploid nucleus from two haploid nuclei.

Haploid nucleus

Oocyte

Sperm

Haploid nucleus

Fertilized egg

Diploid nucleus

How to Pull Genetic Strings

This is one of the truly fundamental questions facing the explorer of the living world, the key to how cells control their fate. For cells change, adapt, differentiate, transform by turning on certain genes and shutting off others. Even the simplest bacteria do this. Yeasts, protozoa, fungi, plants, and animals—all regulate and modulate the activity of their genes. The most notable examples of this are found in the development of higher animals, ourselves included.

Take the human egg cell just before fertilization. It is a large cell and its cytoplasm is stocked with reserves and with organelles ready to get into the act. Its nucleus is haploid (Greek *haplous*, single), which is to say that it contains a single set of 23 chromosomes, totaling about 3 feet of DNA selected from the mother's genetic endowment through the lottery of meiosis and crossing-over (see Chapters 18 and 19). Upon fertilization, it receives a second set of similarly selected chromosomes from another haploid cell, the male spermatozoon, which is little more than a DNA torpedo equipped with a flagellar motor at the back and an invasion apparatus in front. This is the acrosome (Greek *akron*, tip), a modified lysosome. Through fusion of the two nuclei, the fertilized egg cell becomes diploid (Greek *diplous*, double). It now has 46 chromosomes, containing a total length of about 6 feet of DNA, or 5.8×10^9 nucleotide pairs, one out of $10^{3,480,000,000}$ possible combinations, guaranteed to be unique against all conceivable odds. (It would take more than five thousand average-size books just to print that figure: 1 followed by more than 3 billion zeros.) In this string of some 6 billion nucleotide pairs is written the whole predetermined history of one particular individual from conception to death, as well as much of what is environmentally actuated but depends on the kind of response the individual offers to an outside challenge.

Soon after fertilization, the egg cell starts to divide, producing two daughter cells, which themselves divide into four cells, then eight, sixteen, and so on. All these cells have the same 6 feet of DNA, the same identical sequence of 5.8×10^9 nucleotide pairs. Nevertheless, they soon cease to look alike. A polarity sets in within the little bunch of cells that has formed (*morula*, small mulberry, in Latin), and the cells start becoming different from each other, depending on where they are situated in the structure. This process of differentiation continues, leading in a few weeks' time to the development of a characteristic embryo in which a number of different cell types are readily recognized. And so event follows upon event, enacting with remarkable fidelity a script of enormous complexity, to produce at the end of nine months the miracle that is a newborn baby: nerve cells, muscle cells, skin cells, retinal cells, liver cells, and many others, uniquely organized to form a brain that is already sending orders in all directions, eyes that blink in the light and are moved by tiny muscles, a beating heart, lungs that fill with air and expel it in the baby's first cry, a mouth that is already seeking the life-giving nipple, a stomach and intestine that are getting ready to run their first digestive trials on the mother's milk, small arms and legs that flail, miniature fingers and toes that wriggle . . . altogether about 1 trillion (10^{12}) cells, each equipped with the same 6 feet of DNA, but nevertheless doing all sorts of different things and forming all sorts of different associations.

And this is only the beginning of a long saga, which normally will take the individual over a span of some 80 to 100 years, first up from babyhood to childhood, through puberty to adolescence and young adulthood, and then slowly down to middle age, old age, senescence, and death. It is all preordained, written into the 5.8×10^9 nucleotide pairs of the nuclear DNA. The egg cells of a mouse, mare, or chimpanzee do not look very different from those of a woman. Their cytoplasmic machineries are virtually indistinguishable. Nevertheless, each unerringly unfolds its own story, produces its own specific kind of miracle: a newborn mouse, a foal, a baby chimp. They follow different programs; their DNAs are different.

Not only life cycles and basic developmental processes are DNA-dependent in this way. So are many short-term changes and adaptations. Think, for instance, of the female sexual cycles and of all the modifications that accompany pregnancy, delivery, lactation; of the training of athletes; the healing of wounds; the adaptation to climate; the many cycles and rhythms that attune our body functions to the hour of the day. Mind you, not all adaptive changes operate by way of DNA. In general, only slow responses that take hours, if not days, to manifest themselves, do so. Rapid responses, such as are evoked by many hormones or by nerve impulses, rely on metabolic circuits, feedback loops, and other regulatory devices that are built into the existing cytoplasmic machinery (Chapter 14). DNA-dependent changes are those that alter the machinery itself, generally by adding certain proteins to it or by ceasing to supply others.

How all this complexity operates is formally extremely simple. In any given cell, only a small fraction of the genes are turned on—that is, are actively transcribed. This is how cells endowed with the same DNA manage to be different. All they need do is transcribe different parts of their DNA and thereby make different sets of proteins. Cells are, to an overwhelming degree, the structural and functional manifestations of their proteins. The ultimate control of what a cell is to be or do, therefore, belongs to whatever pulls the levers that turn genes on and off inside the nucleus.

This control is itself genetically controlled and forced to follow a precise four-dimensional program. Whatever happens at step *n* of embryological development determines what levers will be pulled in what cells at step $n + 1$. Nevertheless, the program is not written into the nuclear DNA as some sort of wound-up tape that slowly unrolls in irreversible fashion. This has been dramatically demonstrated by experiments in which the nucleus of an unfertilized egg cell has been replaced microsurgically by a nucleus removed from a fully differentiated cell, originating, for instance, from the intestinal wall. Such eggs can develop into embryos that rarely develop further but whose nuclei can be similarly transplanted to produce fully mature animals. Such development occurs in spite of the fact that the originally implanted nucleus has already played out most of its developmental program and closed, in the process, all but one of the multiple avenues that are open in the pluripotential egg-cell nucleus. It has thereby become possible, by serial transplantation of nuclei, to clone an individual—that is, to produce a number of genetically identical copies of the donor of the original nucleus. So far, such experiments have been successfully performed only with amphibians and lower forms of life. In spite of an imaginative novelist's claim, the cloning of mammals, let alone man, has not yet been done, which does not necessarily mean that it cannot be, and will not be one day.

In any case, these experimental achievements do not derive their meaning from the more freakish and sensational applications that they may allow. What they are telling us is a vitally important piece of information: the nucleus of a differentiated cell is not necessarily committed irrevocably to the expression of a single gene set. Its genome can be reawakened, either in the nucleus itself or in nuclei derived from it by mitotic division, by means of messages originating in the cytoplasm. It is obviously the egg-cell cytoplasm that orders the intestinal-cell nucleus or its offspring to re-enact the whole developmental program of the species.

Not all nuclei conserve their pluripotential property in the course of differentiation. As we will see in Chapter 18, maturing lymphocytes offer a striking example of gene processing. Other such cases may well exist. But the very fact that cloning has been achieved should suffice to correct any exaggerated idea we might have formed about

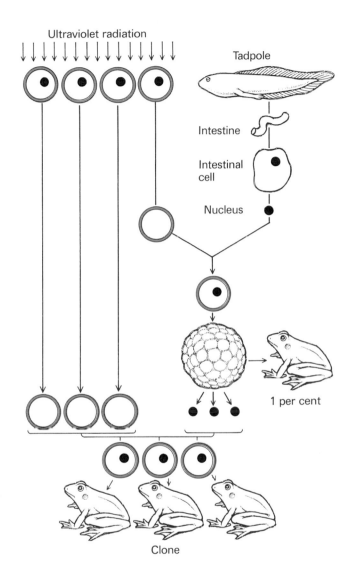

Cloning experiment demonstrating the pluripotential character of the nucleus of a fully differentiated cell. The nucleus of an intestinal cell of a tadpole is implanted into an unfertilized egg cell whose nucleus has been inactivated by ultraviolet irradiation. Some of the renucleated egg cells develop into early embryos (blastulas), of which a small fraction (about 1 per cent) continue their development to the adult frog stage. A second-generation clone of frogs with identical genetic endowment derived from the single original intestinal nucleus can be obtained by similar implantation of a number of nuclei separated from the first-generation embryo.

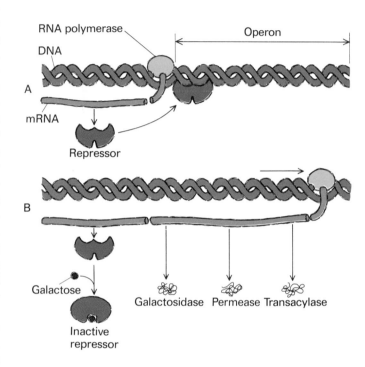

Jacob-Monod model of the lac operon.

A. A repressor prevents the transcription of the operon.

B. Galactose or another inducer binds to the repressor and renders it unable to bind at the promoter site of the operon. Transcription proceeds and translation of the resulting mRNA leads to the synthesis of three enzymes coded by the genes in the operon.

the power of the nucleus. In spite of its central location in the cell and ultimate guardianship of the organism's genetic endowment, the nucleus is not the autocratic despot that we might have suspected. It is not much more than an articulated puppet, admirably constructed and programmed, to be sure, but nevertheless manipulated continuously by the very objects of its control. When a nucleus turns certain genes on or off, it does so in response to orders received from its surrounding cytoplasm, or sometimes from even further afield through messengers produced by other cells, or through drugs, pollutants, or other substances coming from the outside. The cytoplasm, however, is no more boss than is the nucleus; its messages are conveyed or produced by proteins synthesized according to nuclear instructions. In other words, nucleus and cytoplasm do no more than interact with each other in a reciprocally coordinated fashion. The cell is a cybernetic system (Greek *kybernêtês*, steersman). And so, through its superimposed network of cell–cell interactions, is the organism.

As mentioned, the nucleus sends out instructions to the cytoplasm by means of RNA molecules transcribed from certain genes. How, now, does the cytoplasm tell the nucleus what genes to transcribe? Unfortunately, we have a fairly clear answer to this question only for certain elementary bacterial systems, to which the model proposed by the French scientists François Jacob and Jacques Monod is applicable. The main feature of the Jacob-Monod model is the presence of repressor molecules, specific proteins that bind to DNA just downstream of a promoter binding site for RNA polymerase and thereby impede transcription of the particular DNA sequence controlled by this promoter site. In bacteria, such a sequence often includes several genes that are transcribed into a single polycistronic mRNA. Such a set of genes is called an operon. Activation—or, more precisely, derepression—of this set of genes is effected by small molecules, called inducers, that combine with the repressor and thereby modify its conformation in such a way as to cause it to detach from the DNA. The first operon to be recognized was the lactose operon of the bacterium *E. coli*. Bacteria grown in the absence of lactose (milk sugar)

lack the ability to utilize this sugar, for want of some key enzymes. But if they are exposed to lactose, they soon acquire the necessary enzymes and thrive on the new substrate. It took the French workers, together with many other investigators, some fifteen years of arduous and ingenious experimentation to discover that this remarkable instance of metabolic adaptation is explained by the ability of lactose to detach from the DNA a bound factor of protein nature (the repressor) that blocks transcription of a set of three genes (the lactose operon) coding for three enzymes of lactose metabolism.

Since these celebrated experiments were performed, a number of other operons have been discovered and much detail has been added to our understanding of their control. Unfortunately, these interesting bacterial systems may not tell us too much about the manner in which eukaryotic genes are regulated. So far, no operon has been identified in a eukaryotic nucleus; neither has a typical

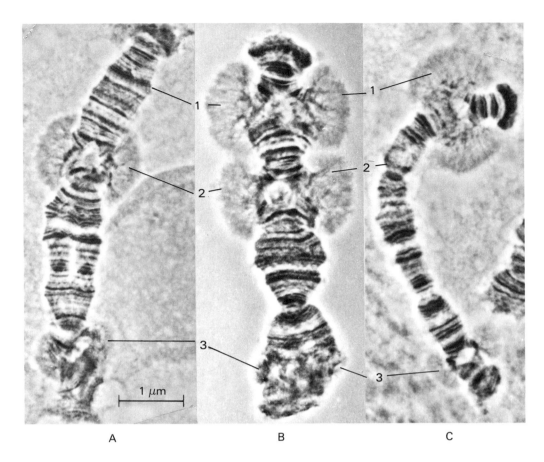

Giant RNA puffs, known as Balbiani rings, on polytene chromosome IV from salivary gland of *Chironomus tentans.*

A. Animal in natural habitat. Region 1 is unexpressed; regions 2 and 3 are fully expressed.

B. Larva exposed to high titer of ecdysone. Regions 1 and 2 are fully expressed; region 3 is poorly or not expressed.

C. Larva fed on galactose. Regions 1 and 3 are fully expressed; region 2 is completely repressed.

repressor protein. On the other hand, some information specifically relevant to eukaryotes has been uncovered, largely through the study of thyroxin and especially of steroid hormones.

The steroids are a family of molecules derived from cholesterol. They include several hormones that act on developmental processes. For instance, androgens control the production of sperm and the expression of various male characters such as the growth of facial hair; estrogens regulate the menstrual cycle and induce such female characters as the development of mammary glands. It has been well established that these effects are due largely to the production of specific proteins by the target cells, a phenomenon which is itself the consequence of the issue of specific mRNAs out of their nuclei. Steroids, in other words, act by turning on certain genes. How they do this is known in roughly schematic fashion. In contrast with most other hormones, which bind to surface receptors,

the steroid hormones, being small and strongly hydrophobic, cross the plasma membrane readily and are picked up in the cytosol by soluble receptor proteins. The hormone-receptor complex then moves into the nucleus through one of the nuclear pores and binds to certain parts of heterochromatin, causing them to unroll into actively transcribed euchromatin. Unless you have a means of taking very precise bearings inside the nucleus, you may have difficulties in recognizing this change. But certain insects offer a made-to-order situation in which the arrangement of chromatin is easily perceived, even by rudimentary means. It occurs in the giant polytene (Greek for multistranded) chromosomes found in the salivary-gland cells of diptera larvae. In these cells, chromosomal fibers are duplicated manifold (polyploidy), up to thousands of times, outside of any mitotic activity. Identical fibers are accurately aligned against each other, forming thick, extended chromosomes in which individual genes, or small sets of genes, appear as alternating bands of variable density. As mentioned earlier in this chapter, the Morgan school actually located a large number of genes on these chromosomes. When such larvae are treated with ecdysone (Greek *ekdyein*, to unclothe), a steroid hormone that induces molting, certain condensed bands of the polytene chromosomes are seen to blow up into typical "puffs," which can be shown by special techniques to be centers of active mRNA synthesis.

So much is clear. Not so the mechanism whereby hormones or hormone-receptor complexes disperse DNA fibers and stimulate their transcription in certain selected areas of the genome. One of the prerequisites of transcription is the uncoiling of the chromatin threads and loosening of the nucleosomes. Understandably, many cell explorers are at present scrutinizing the histones around which nucleosomes are built, looking for changes, such as acetylation or phosphorylation, that might underlie the structural changes that accompany transcription. But all nucleosomes contain the same set of histones. There must, therefore, be sites on the DNA itself that are recognized by the hormones or their receptors in such a manner as to attract and put to work the loosening enzymes and RNA polymerase.

The selection of certain genes for transcription is only the first step in a process that also involves extensive subsequent editing of the primary transcripts, sometimes in more than one way. No doubt a number of additional controls are exerted at this level. Some explorers even claim these to be the main controls. They believe gene expression in eukaryotes to be regulated, not so much transcriptionally, as it is in bacteria, but posttranscriptionally, through selection of the messages that are to be edited and issued to the cytoplasm. Furthermore, as we shall see in Chapter 18, extensive pretranscriptional editing of genes may also occur, though probably in only a few exceptional instances.

A remarkable example of genetic regulation of a particularly drastic kind takes place in response to fever, and perhaps other stressful situations. Known as the "heat-shock" phenomenon, it is characterized by the sudden rapid synthesis of a set of five or six proteins, with the concomitant inhibition of the formation of most others. According to some reports, temperature may act by dismantling the RNP complexes that are involved in RNA splicing; pre-mRNA would continue to be made at close-to-normal rate but would no longer be processed and transferred to the cytoplasm. Transcripts that contain no introns would continue to be delivered and translated, and would be so even at a considerably faster rate, perhaps because they appropriate the protein-synthesizing machinery left idle by the slowing down of mRNA production. The resulting "heat-shock proteins" are believed to play a role in the defense against stress.

17 | DNA Replication and Repair

While we were struggling through the dense nuclear underbrush and seeing its tangle of fern fronds, lianas, and other strange molecular vegetation turn into thousands of transcription centers, all busily copying and processing instructions for delivery to the cytoplasm, a momentous decision was being weighed in this very cytoplasm: to divide or not to divide?

A Fateful Commitment

The question confronts every daughter cell after it has emerged from the so-called M phase (for <u>m</u>itosis) and entered the G_1 (gap <u>1</u>) phase of its cycle. If the answer is yes, the order is relayed to the nucleus, where preparations are started for duplication of its information content. Within a few hours, the cell enters the S phase (for DNA <u>s</u>ynthesis), which can be conveniently detected by timing the moment when the cell incorporates added radioactive thymidine into DNA. From 6 to 10 hours later, the entire genome has been replicated and the cell goes into the G_2 (gap <u>2</u>) phase, where it stays for the time it needs to get ready for mitotic division (2–6 hours). Mitosis itself takes about 1 hour. The length of the complete cycle, for the cells of higher animals, is of the order of 20 to 24 hours. Bacteria, which have no nucleus and a much simpler life cycle, may start duplicating their DNA even before separating from their twins. Their doubling time can be as short as 30 minutes.

The rapid growth of bacteria is of great practical importance in both laboratory research and industry. Given

The cell cycle. After its birth by mitotic division (M), a cell enters the Gap-1 (G_1) phase, in which it may remain indefinitely (G_0). Commitment to division occurs at some stage of the G_1 (G_0) phase, leading after a few hours to duplication of the DNA load of the nucleus by new synthesis (S). A short Gap-2 (G_2) phase separates the end of DNA duplication from the onset of cell division. The approximate duration of each phase in a 24-hour cycle is 1 hour for M, 11 hours for G_1, 8 hours for S, and 4 hours for G_2.

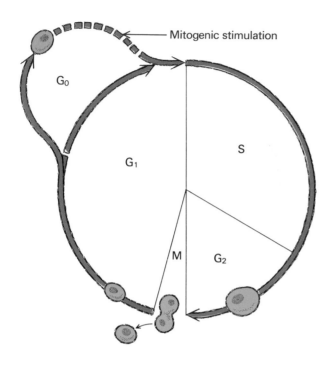

Mitogenic stimulation

G_0

G_1

S

M

G_2

What orders this all-important shunting is still very poorly understood. Crowding is one factor. Fibroblasts cultured on a flat dish stop growing when they reach confluence. Somehow, when the cells touch each other (contact inhibition) or reach a certain density (density-dependent inhibition), surface receptors are activated that switch the cells into G_0 phase, perhaps with the help of cyclic AMP. Something similar occurs in vivo. If one removes part of the liver from a mouse, the remaining liver cells immediately resume their growth cycle. They go on dividing until the original weight of the organ has been restored, and then they return to G_0. Similarly, after nephrectomy, the remaining kidney undergoes "compensatory hypertrophy." These phenomena, however, can hardly be explained simply on the basis of contact inhibition or lack of it, since the cells appear to sense at a distance the total number of their congeners that are present and to switch from G_1 to G_0 or back according to the information they receive. Presumably, some humoral factor is involved. The discovery of cell-specific growth factors is beginning to shed light on this central problem of growth regulation.

These examples, together with many other known instances of mitogenic stimulation, make it clear that the switch to G_0 is not irrevocable. Cells can be reawakened from their quiescent state by a variety of chemical calls. Many important physiological mechanisms depend on this property, which, however, will serve its purpose only if the return to G_0 can again be enforced stringently at some appropriate time. Let a single cell discover a means of escaping this constraint and of passing on the secret to its progeny, and an ineluctably fatal exponential growth process will be initiated, even if the loophole is of such poor quality that one might be tempted to brush it off as of no consequence. Imagine, for instance, that a cell has found an inheritable way of getting back illegally into G_1 only once in every hundred days. No cause for alarm, you might say. Indeed, it will take the lawbreaker's offspring five years to reach the tiny weight of one milligram and eight years to grow to the size of a small pea that might possibly be detected if favorably situated. After that, however, things suddenly start moving precipi-

an adequate supply of nutrients, a single bacterial cell can generate 280,000 billion ($2.8 \times 10^{14} = 2^{48}$) individuals in a single day. During that time, the average human cell has barely doubled. But don't let the word "barely" delude you. If allowed unrestricted, exponential multiplication, such a cell could produce the equivalent of an adult human body in 6 weeks and of the volume of our planet in just about 4 months. Even an embryo does not grow that fast; in fact, the net growth rate of the developing organism decreases steadily until it becomes essentially zero in the adult. Cell death accounts only minimally for this limit to growth. With the exception of a few cell types—primarily those of blood, skin, and mucosae—most cell populations in the adult organism turn over very slowly or not at all (in contrast to their constituents, which we have seen turn over at appreciable rates). This is because cells have the ability to interrupt their growth cycle and to remain indefinitely in the G_1 phase, in which case they are said to be in G_0 phase.

tously, leaving very little time to act. Two years later, the pea has turned into a one-pound tumor, which would itself require only another two years to reach the size of the body were not its growth interrupted by the death of its exhausted host. It is the whole drama of cancer. No wonder so many cell explorers now focus their scrutiny on the fateful G_0-G_1 switch.

What long-range future lies in store for the particular cell that we are visiting is hidden from us. But its immediate fate is clear. Several unmistakable signs tell us that the cytoplasm has reached, and relayed to the nucleus, the decision to move full steam ahead from G_1 into S. Cohorts of unfamiliar proteins are infiltrating the nuclear sap. And everywhere, even in the darkest corners of dormant heterochromatin, faint tremors are beginning to stir the curled-up DNA threads. It is obvious that some portentous event is about to take place.

The Secret of Faultless Copying

The event we are about to witness, as no doubt you have guessed, is DNA replication. To make sure of a front seat, let us move to one of those points where a chromatin fiber is anchored to the nuclear matrix. It is an irregularly shaped knob, made of a number of different proteins. As we watch it, this structure is clearly undergoing some sort of reorganization, apparently triggered by the advent of new proteins. Suddenly, it comes alive; it starts reeling in its attached chromatin arms from both sides and extruding them through its center in the form of a widening loop. You might well wonder at the utility of this molecular cranking until you look more closely at the chromatin loop that is coming out. It consists not of one loop, but of twin loops, identical in every respect. The machine we are watching is not a simple translocator; it is a replicator, sometimes called a replisome.

While this is going on, neighboring anchoring points have also turned into replisomes. In less than 1 hour, the long chromatin stretches strung between them have been completely dragged in and replaced by the duplicated loops that come out of their centers. In the end, adjacent replisomes, drawn together by their opposed pulls on the tightened chromatin between them, mingle in a final convulsion to join their products into a continuous duplex. Up to several thousand replisomes operate in this way all along each of the 46 chromatin fibers. They work as clusters of from 25 to 100 replisomes, but the different clusters are not all synchronized. Some start early in the S phase, others get into action late, so that it takes them a few hours altogether to finish their job. By that time, each chromatin fiber has been completely duplicated from end to end. This means not only replicating the DNA, but also doubling the number of nucleosomes with the help of new histone molecules provided by the cytoplasm, constructing a second matrix framework, and redistributing the two daughter chromatin fibers around the scaffolds in a way that will allow their subsequent separation as distinct chromosomes. Many of these remarkable events are known to us only through their outcome. They are hidden in the darkness of impenetrable tangles, like the secret life of a jungle, and no explorer has yet succeeded in setting eyes on them. But there is one important exception. Replisomes have been opened and dissected into their main component parts. We now have a fairly good idea of what goes on inside them.

Basically, DNA replication takes place according to the same tail-growth mechanism as does its transcription, described in Chapter 16. Deoxynucleotidyl groups are transferred from the corresponding dNTPs to the 3′-terminal hydroxyl group of the growing chain, the choice among dAMP, dGMP, dTTP, and dCTP being dictated by the nature of the base—T, C, A, or G—facing assembly on an antiparallel DNA strand that serves as guide. Besides these similarities, a number of notable differences distinguish replication from transcription. The most important ones are:

1. Replication affects both strands of the DNA duplex.

2. The product of replication remains associated with the instructing strand as a duplex consisting of one parental and one daughter strand (semiconservative replication).

3. Replication affects the totality of the genome.

4. Replication is several orders of magnitude more accurate than transcription.

5. Unlike RNA polymerase, DNA polymerase, the main agent of replication, cannot start anew with a dNTP as first acceptor; it needs a primer, a short, preformed DNA or RNA chain.

The unit of replication is called a replicon. It represents the entire stretch of DNA that is replicated by a single replisome, starting bidirectionally from the midpoint of the replicon. This central initiation point is called the origin of replication. Bacterial and viral DNAs consist of single replicons, which, in bacteria, may exceed 1 mm in length. In contrast, eukaryotic chromosomes contain hundreds of replicons, sometimes as many as several thousand. Their sizes vary from 50,000 to 300,000 base pairs (from about 15 to 100 μm of DNA, or 0.3 to 2 μm of chromatin).

If the view of replication that we have just been granted is correct—remember, facts are still scanty and imagination much in demand—then each chromatin region attached to the nuclear matrix is a replication origin, and each intervening chromatin segment is made of the halves of two contiguous replicons. At least, this is the disposition when the replisomes have been completed and replication is set in motion. So far, we know the structure of the replication origin only for relatively simple viral systems. The impression gained from such systems, in both the prokaryotic and the eukaryotic worlds, is that certain "palindromic" structures play an important role in initiation. (A palindrome is a word or phrase that reads identically in both directions). For instance, in the small animal virus SV40 (p. 312), the origin of replication contains the following sequence of 27 base pairs, which, except for the central G—C pair, reads identically as written or turned around 180°:

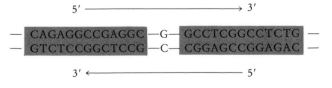

This kind of structure makes sense for a process that is expected to start symmetrically on the two strands and to share their antiparallel polarity. Furthermore, it is also capable of rearranging into opposed "hairpin loops," or cruciform structure, which may help in initiation:

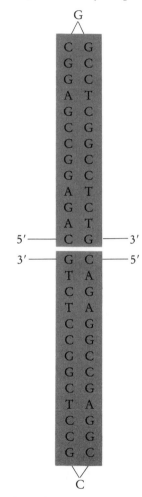

Initiation involves the participation of special proteins known as initiators. It brings into play two distinct events: strand separation and priming. Strand separation requires uncoiling a stretch of DNA at the expense of compensatory supercoiling at both ends. There are enzymes, called helicases and gyrases, that can wring DNA

DNA replication.

1. A replisome has assembled at an anchoring point of chromatin loops on the nuclear matrix (the spooling of DNA in nucleosomes is not shown for simplicity's sake). After strand separation by topoenzymes, replication is initiated in both directions by two primases, each of which assembles the RNA primer of a leading strand and opens a replication fork. The DNA of the replicon controlled by the replisome begins to be reeled in from both sides. A bubble forms between the two strands.

2. DNA polymerases have replaced the primases and started the synthesis of DNA on the leading strands. The bubble widens as more DNA is reeled in.

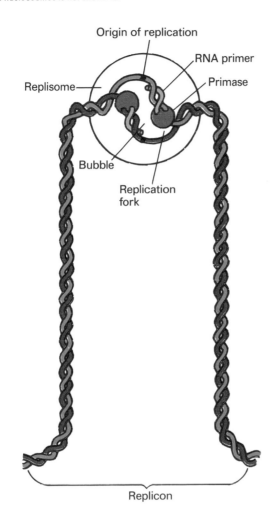

1

Origin of replication

RNA primer

Primase

Replisome

Bubble

Replication fork

Replicon

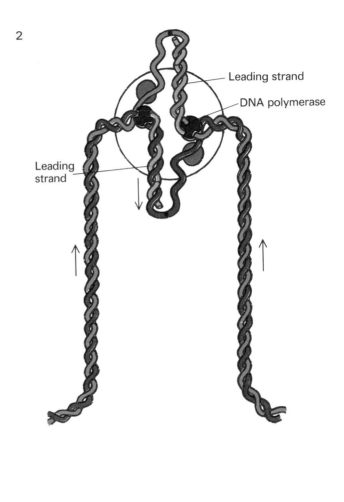

2

Leading strand

DNA polymerase

Leading strand

forcibly in one or the other direction with the help of ATP. Certain DNA-binding proteins, called unwinding, melting, or helix-destabilizing proteins, can do the same cooperatively by linking up all along each strand. Finally, stresses in DNA can be relieved by topoisomerases (nicking-closing enzymes), which have the ability of nicking one of the DNA strands, while conserving the bond energy, and of closing the bond again after the supercoiled duplex has relaxed by rotating around the swivel provided by the unnicked strand. We saw these enzymes at work to help transcription. Some topoiso-

merases even manage the extraordinary trick of nicking both strands, letting a loop pass through the opening, and closing the gap correctly afterward.

Priming is carried out by a special kind of RNA polymerase, called primase, that builds a short stretch of complementary RNA on each DNA strand, with ATP as obligatory starting acceptor at the 5' end. Only nine or ten nucleotides long, on average, this priming RNA stretch plays an essential role because, unlike RNA polymerase, DNA polymerase cannot start a DNA chain de novo. The priming RNA is removed later.

As soon as the priming RNA has been assembled, DNA polymerase takes over and actual replication starts. In this process, the polymerase cuts through the DNA duplex, much as does the RNA polymerase, with the same topological restrictions that require it to be preceded by a nicking enzyme to allow rotation of the template strand. The term replication fork designates the Y-shaped structure at the leading edge of this movement, where the two strands separate from the duplex. Because of the characteristic symmetry of the replication origin and of the replisome, these events take place bidirectionally and generate two opposed replication forks, with the help of two primases and subsequently of two polymerases, assisted by the indispensable topoisomerases.

In transcription, as we saw in Chapter 16, the polymerase—more often a train of polymerases—courses along the DNA following the instructing strand in the 3′——→5′ direction; the disjoined DNA closes up directly behind the advancing enzyme. Things are different in replication. Each of the two polymerases remains a single and, as far as we know, stationary part of the matrix-bound replisome complex. It is the DNA that does the actual moving, under the pull of the replication process. Relative displacement and dynamics are the same in replication and in transcription. It is simply a question of what moves and what stands still with respect to the nuclear framework.

A more fundamental difference is that the separated DNA strands do not rejoin behind the polymerase, and the gap between them therefore widens to a point at which it can be easily seen in the electron microscope. Called a bubble, or eye, this gap is symmetrical in shape and limited at each end by an advancing replication fork. But watch out: the symmetry is not perfect. Each fork has a thick and a thin branch. The thick branch is the semiconservative duplex between daughter DNA and parental template; the thin branch is the other parental strand, as yet unreplicated. This is because polymerases are one-way enzymes. They can read only in the 3′——→5′ direction and assemble in the 5′——→3′ direction. For the same reason, the asymmetry of the forks is inverted: the thick branch of one is continuous with the thin branch of the other, and vice versa, because each polymerase is replicat-

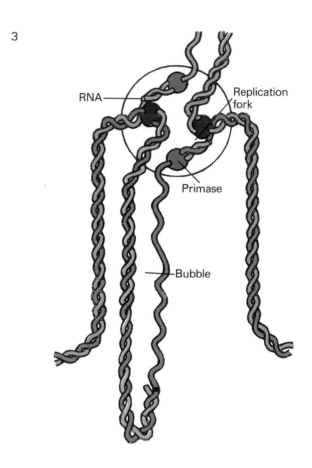

3

RNA

Replication fork

Primase

Bubble

ing a different strand, consistent with the direction of its movement.

This mechanism could conceivably suffice in the case of circular DNA if the two polymerases were somehow allowed to complete a full circle along their respective template strands until they bump into the 5′ tails of their own primers. In practice, however, this does not happen, even with short circular DNAs. The thin strands left behind by the two leading polymerases are replicated separately by other primase-polymerase systems. Owing to the obligatory polarity of assembly, this process has to move backward from the direction in which a single-

4. DNA polymerases have replaced the primases and have started the synthesis of DNA on the lagging strands. The arrows show the direction of movement of the template

DNA (pink) as it is being reeled in through the replisome.

5. DNA polymerases have almost completed the first Okazaki fragments

of the lagging strands and are beginning to displace the RNA-primer ends of the leading strands. The displaced RNA is broken down by 5′-3′-exoribonucleases.

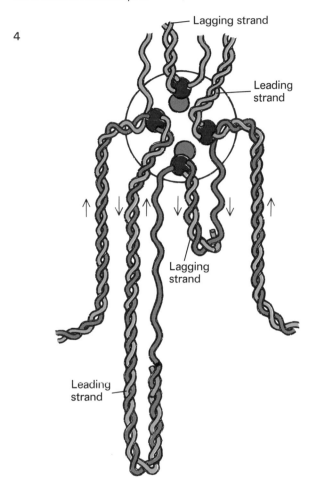

4

Lagging strand

Leading strand

Lagging strand

Leading strand

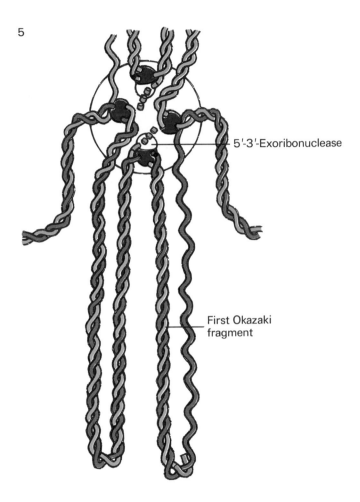

5

5′-3′-Exoribonuclease

First Okazaki fragment

stranded template is made available by the advancing replicating forks. To meet this requirement, replication of the thin strand takes place by sudden forward leaps of the primase system, followed by backward, gap-filling elongation, to produce a series of short DNA stretches that will have to be stitched together later. Thus, whereas each strand is replicated continuously in the 3′⟶5′ direction from the point of origin, it is replicated discontinuously in the other direction. The whole process is called semidiscontinuous. Because discontinuous replication starts only after a certain length of single-stranded DNA has been liberated by the continuous process, the strand it

generates is called the lagging strand. That which is synthesized continuously is called the leading strand. The terms forward arm and retrograde arm are also used.

Discontinuous replication functions by the same mechanism as does continuous replication. Primase starts the process and DNA polymerase then takes over until it runs into the 5′ end of another primer. Without further action at this stage, the product consists of a series of discrete DNA fragments (with a short 5′-RNA tail) coiled around the template strand. These fragments are called Okazaki fragments, from the name of the Japanese investigator who discovered them (and later died of leukemia: he had

6. The first Okazaki fragments of the lagging strands are completed and the RNA primers of the leading strands have been removed. Ligases now take over and attach the 3′ end of each

Okazaki fragment to the 5′ end of each leading strand.

7. Primases have started the second Okazaki fragments of the lagging strands, while the leading strands

(with first Okazaki fragments attached to their 5′ ends) continue to grow. Steps 4 through 7 are then repeated as many times as is required to reel in the DNA of the replicon completely.

6

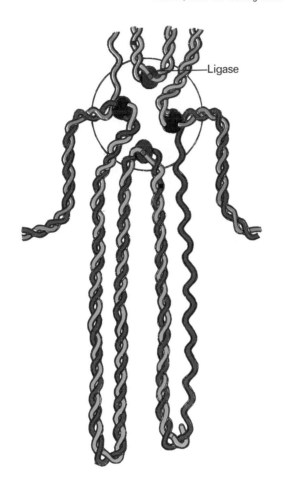

7

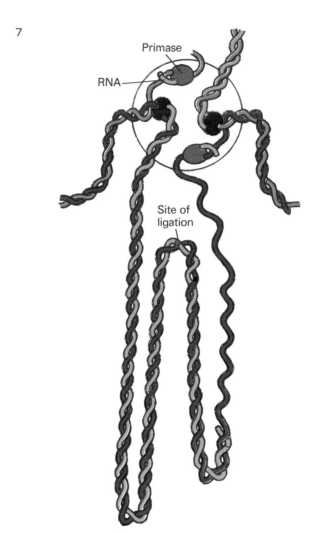

the misfortune as a child to grow up in Hiroshima). These fragments are some 1,000 to 2,000 bases long in bacteria and only some 180 to 280 bases long in eukaryotic cells. The latter periodicity is believed to be related to the organization of eukaryotic DNA into nucleosomes. It certainly indicates that primase is not finicky as concerns its attachment site when presented with a single DNA strand.

Completion of the discontinuous process requires the excision of the RNA primer that blocks the further progress of the DNA polymerase, the pursuit of replication

until the 5′ end of authentic DNA is reached, and, finally, the splicing of this 5′ end with the leading 3′ end of the elongating fragment. The first step is catalyzed hydrolytically by a special 5′-3′-exoribonuclease (so named because it attacks RNA at its 5′ end, removing 3′-mononucleotides in succession). In bacteria, this activity is an intrinsic property of the main DNA polymerase and is situated at the front of that enzyme so as to clear the way automatically for replication to proceed. The final splicing step is catalyzed by an AMP-forming ligase (Chapter

8. When replication of two adjacent replicons is almost completed, the DNA polymerases that elongate converging leading strands are drawn together by the shortening DNA loops between them and start to reel in the single-stranded DNA supporting the neighboring lagging strands.

9. The RNA primers of the last Okazaki fragments of the lagging strands are displaced and broken down as DNA synthesis proceeds to completion.

10. Ligases join the fully replicated neighboring replicons.

8

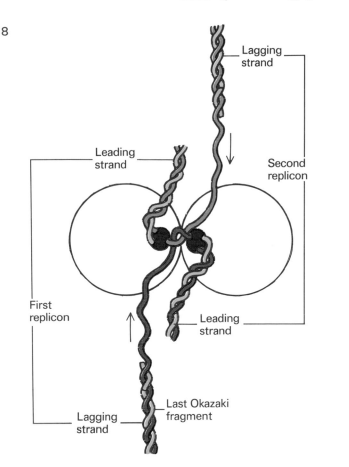

9

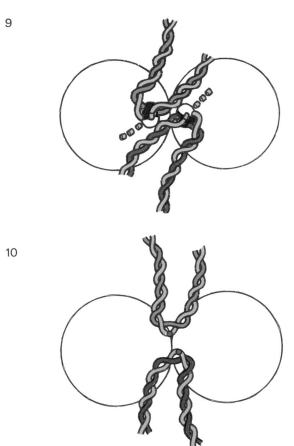

10

8). This enzyme activates the 5′-phosphoryl end-group by AMP-yl transfer to make an adenylyl-DNA Janus intermediate, which is then attacked on the other side of its central oxygen atom by the free 3′-hydroxyl of the joining partner, with liberation of AMP.

In circular bacterial chromosomes, the two forks meet about halfway round. Their orientation is such that the 3′ head of each leading strand encounters the 5′ tail of the corresponding lagging strand, and vice versa. These ends need only to be joined by the mechanism just described to produce two identical, semiconservative, circular duplexes. Such rings are sometimes linked and must be disentangled by special topoisomerases. In eukaryotic chromosomes, exactly the same kind of head-to-tail joining takes place at the point of meeting between two forks that belong to adjacent replicons. Replication of the whole chromosome is completed when all its replicons are joined. Such are the basic mechanisms of DNA replication. There are many variants, especially in the viral world, but we will not go into those.

Schematic view of how the multiple bubbles shown in the electron micrograph on the facing page are believed to be looped on matrix-affixed replisomes. Unduplicated DNA is shown in blue, duplicated DNA (bubbles) in red.

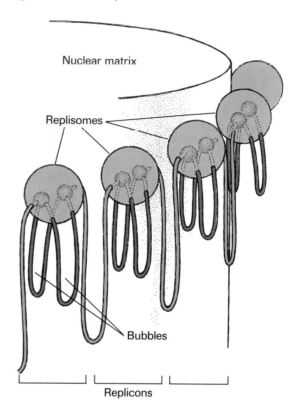

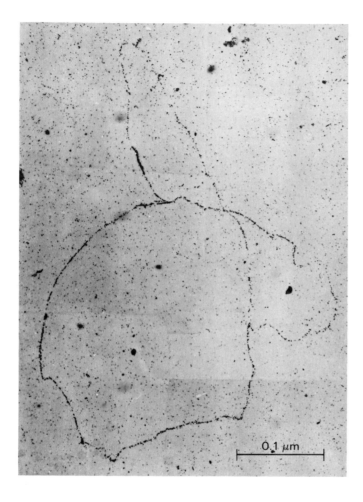

Image of a bacterial chromosome that has been detached intact from the replisome while in the process of duplicating. The two replicating forks are clearly visible. The bubble is twisted into a figure of eight. This picture was obtained by autoradiography. The bacteria were grown on a medium containing radioactive ³H-thymidine to label the DNA. The cells were then killed, and their DNA was extracted by a gentle procedure designed to minimize breakage of the fragile molecules and was spread in close contact with a sensitive unexposed photographic film. Radiation emitted by the DNA impressed the film, tracing the shape of the molecule.

The problem of accuracy does, however, deserve our attention. As mentioned in the preceding chapter, RNA polymerase has an error frequency of the order of 10^{-4}. This is tolerable, considering that the mistakes are likely to be randomly distributed and that many of the resulting base substitutions will not be expressed in functionally defective proteins (owing to the properties of the genetic code or because the substitutions occur at strategically unimportant points in RNA stretches that are not translated). Therefore, transcription mistakes may cause some waste but hardly ever any serious harm. The situation is entirely different for replication. There, the mistakes are made in a master copy, whose defects will be imparted to all of its RNA transcripts and to all of its DNA progeny. They are real mutations. Their consequences may not always be dramatic. But the probability of their being harmful is such that no cell line could survive very long with a 10^{-4} frequency of replication mistakes. Yet, DNA

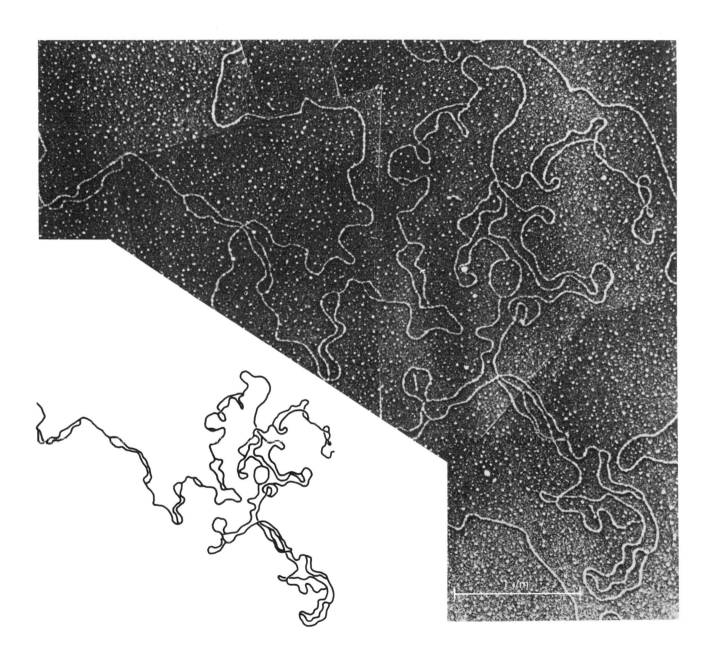

This electron micrograph shows multiple bubbles on a length of eukaryotic DNA (from the fruitfly *Drosophila melanogaster*) in the process of duplicating. The DNA has been extracted, spread on a surface, and revealed by shadowing with metal vapor.

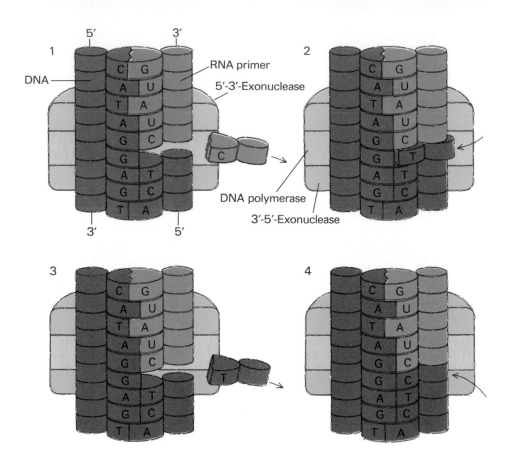

E. coli DNA polymerase I, an enzymic combine shown in action: (1) 5'-3'-exoribonuclease situated in front removes CMP from the 5' end of the RNA primer; (2) polymerase situated in the middle erroneously attaches a dTMP group to the 3' end of the growing DNA chain; (3) 3'-5'-exonuclease situated at the back splits off mismatched dTMP; (4) polymerase correctly inserts a dCMP group.

Labels in figure:

1 — 5' ... 3' ... DNA ... RNA primer ... 5'-3'-Exonuclease ... 3' ... 5'

2 — DNA polymerase ... 3'-5'-Exonuclease

polymerase in itself does not seem to be more accurate than RNA polymerase.

The answer is proofreading, an operation carried out mainly by a special 3'-5'-exodeoxyribonuclease that has the ability to recognize a mismatched base at the 3' end of a growing DNA chain, and to excise the wrong mononucleotide before it is covered by the next one. Then polymerase can try again, with 9,999 chances in 10,000 of not repeating the mistake. Thanks to this and a few other accessory safeguards, replication mistakes are reduced to a level of some 10^{-8} to 10^{-10}. This is obviously tolerable, but it is far from negligible in regard, for instance, to the number of nucleotides (about 1.2×10^{10}) that are assembled every time a genome the size of the human genome is duplicated.

In bacteria, proofreading is actually carried out by the DNA polymerase itself, which possesses an incorporated 3'-5'-exonuclease activity placed so as to "inspect"—and remove if necessary—each nucleotidyl group immediately after its transfer. This same enzyme, you may remember, also bears on its front end the 5'-3'-exonuclease activity needed to clear away any obstructing RNA primer. So far, no such remarkably coordinated molecular "combine" has been spotted in eukaryotic cells. But this may be simply because its pieces fall apart more easily upon isolation. Certainly there is no reason to believe that the efficiency and fidelity of replication are of lower quality in eukaryotes than in prokaryotes.

Considering the many steps involved, the speed of DNA replication is remarkable. In eukaryotes, it proceeds at the rate of about 1 μm per minute, comparable to that of transcription. In prokaryotes, it can be as much as 30 times as fast, which means the addition of 1,500 deoxynucleotides per second and rotation of the template DNA at the fantastic speed of 9,000 revolutions per minute. Presumably, eukaryotes owe their slower rate of DNA replication largely to the organization of their DNA into nucleosomes. However, they more than compensate for this drawback by the smaller sizes of their replicons and Okazaki fragments, which allow a much

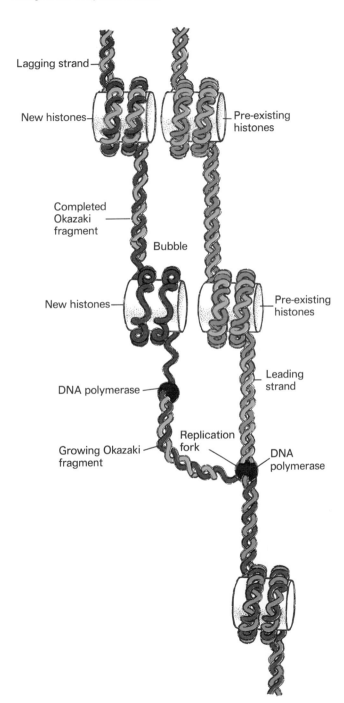

Except for the necessary loosening of their structure, nucleosomes are not dismantled during replication. Pre-existing histone spools remain with the leading strand. Newly made histones form spools for the lagging strand. Whether they bind the strand before replication (as shown in the diagram) or after it is not known.

Lagging strand

New histones

Pre-existing histones

Completed Okazaki fragment

Bubble

New histones

Pre-existing histones

DNA polymerase

Leading strand

Growing Okazaki fragment

Replication fork

DNA polymerase

larger number of replicating systems to be put to work on a given length of DNA. In consequence, a human cell will take only ten times as long as does a bacterium to duplicate a genome that is 2,000 times the size of the bacterial genome.

We are unable to appreciate what additional topological problems the nucleosomal organization creates in eukaryotes or how they are solved. But one thing seems clear. There is no sharing of histones between new and old nucleosomes. Apparently, the leading strand hangs on to the pre-existing histone cushions, and newly made histones sent in from the cytoplasm join the lagging strand. The similarity between the length of DNA per nucleosome and that of the Okazaki fragments suggests that priming on the lagging strand may be related in some way to nucleosome spacing.

DNA Maintenance and Repair

They are the silent patrol, and they never sleep. Day and night, as much in the broad expanses of nucleoli and euchromatin as in the densest recesses of heterochromatin, they remain tirelessly on the beat, looking for flaws, possible mistakes, accidental blemishes in the cell's irreplaceable store of genetic information, ready to pounce on the slightest hint of ambiguity to set the record straight. They are the cell's DNA repair crew.

As often happens, we first learned of the existence and importance of this invaluable service from the consequences of its breakdown. There is a rare group of patients to whom sunlight is fatal. They suffer from a genetic disease called xeroderma pigmentosum, in which the skin (Greek *derma*), when exposed to ultraviolet light, develops pigmented hardenings (Greek *xêros*) that transform into cancerous lesions. Overexposure to sunlight (watch out, fanatics of the bronzing cult) can produce similar lesions in normal individuals. What happens in these sun victims is a progressive deterioration of their skin DNA due to ultraviolet irradiation injuries. The cause of the disease is not, as was first thought, hypersensitivity to UV rays, but rather faulty repair of lesions that

Excision-repair mechanism.

1. A special excision endonuclease recognizes a lesion and cuts the affected strand on the 5′-end side of the lesion, creating a free 3′ hydroxyl group.

2. DNA polymerase lengthens the nicked strand and displaces the segment bearing the lesion.

3. The displaced piece bearing the lesion is excised by 5′-3′-exonuclease action.

4. Ligase replaces DNA polymerase and stitches the 3′ end of the new segment to the 5′ end of the old strand.

5. DNA is repaired, with a short stretch of newly made DNA replacing the piece bearing the lesion. This stretch could be recognized if a radioactive precursor (for instance, ^{3}H-thymidine) had been offered to the cell between stages 1 and 3: it would be "hot," as opposed to the rest of the DNA which would be "cold." The extent of DNA repair going on in a resting nucleus (in which no duplication occurs) can thus be gauged by the amount of radioactivity incorporated.

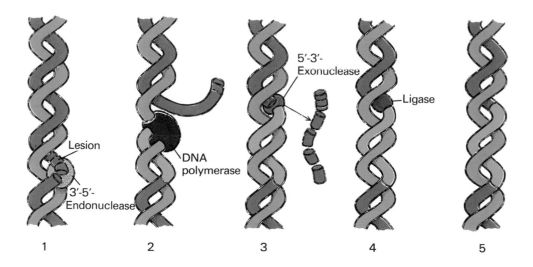

1   2   3   4   5

are not in themselves abnormally severe. Similar deficiencies have been observed and investigated in detail in certain bacterial mutants.

UV light is but one of the many agents that can alter DNA. X-rays, radioactive emissions, free radicals, many different kinds of chemicals, certain enzymes, and, as we have seen, replication errors that escape proofreading—all participate in what is really a constant assault on the genetic orthodoxy of living cells. Modern technology has added a sizeable contingent to this attacking army, but Nature is not as benevolently innocuous as some of our Dr. Panglosses would like us to believe. If it were, cells would not have developed the multiple defenses that they have, and, of course, the evolutionary lottery could never have taken place. The lesions that DNA may suffer are of many different kinds. Strands may be broken or bases removed by hydrolysis. Alternatively, bases may be chemically damaged or mismatched or joined covalently, either directly—the main effect of UV light is to cause dimerization of thymine—or by means of cross-linking agents. So-called intercalating substances may slip in between adjacent base pairs and dislocate the DNA helix.

As is to be expected, the nature of the repair mechanisms depends on the nature of the lesion. Sometimes relatively simple correction of the damage is possible. For instance, a hydrolytic nick is easily repaired by ligase action. Or sunlight may be coaxed into undoing its own damage with the help of a special photoreactivating enzyme that disjoins the UV-induced thymine dimers. In most cases, however, the lesion is irreversible and has to be removed. This is done by glycosylases that split off altered bases or by endonucleases (attacking bonds inside the DNA chain) that cut out a small oligonucleotide segment containing the adulterated part. A new, correct sequence is assembled by a polymerase, and its attachment is completed by a ligase, as in DNA replication.

For such an excision-repair mechanism to function correctly, the enzymes that do the surgery must be capa-

Correlation between life-span and intensity of DNA repair, as assessed by the speed of incorporation of radioactive thymidine into the DNA of resting nuclei in cells that have been exposed to UV irradiation: (left) in different mammalian species; (right) in different primate species.

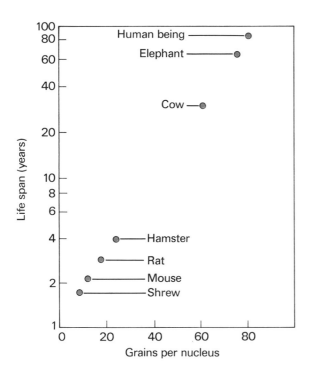

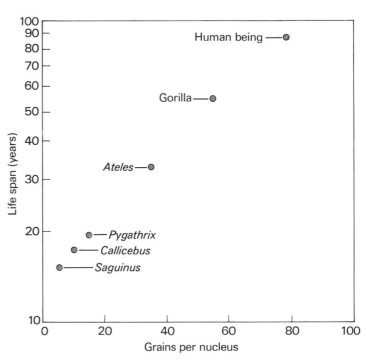

ble of recognizing the lesion. If one strand is obviously abnormal and the other normal, one can visualize, at least theoretically, how such recognition may take place. But what if the two strands are normal but simply mismatched—say, a thymine paired with a guanine? How is the enzyme to decide which of the two is wrong? Indeed, there are cases of genuine ambiguity, where repair has only a 50 per cent chance of being right. But it also happens that the excision enzymes are provided with secondary clues. Methylation is one. If one strand is methylated in the neighborhood of the mismatch and the other is not, the unmethylated strand is identified as the younger of the two and therefore most likely the faulty one, a product of incorrect assembly overlooked by the proofreading system. Excision enzymes are programmed to favor unmethylated strands.

A second prerequisite of correct repair is that the genetic information not be irretrievably lost. One strand must retain the authentic base sequence, correctly associated with the faulty strand, so as to give the right instructions to the replicating polymerase. For this reason, a double nick at the same level is generally fatal. Also, single strands in replicating forks may be difficult to repair. Even lesions of this sort may occasionally be repaired correctly if adequately splinted—for instance, by DNA-binding proteins. You may remember that certain topoisomerases cut a double strand in such a way that a DNA loop can actually pass through the cut, and yet are able to repair the damage correctly afterwards.

Although not infallible, our DNA repair crew is certainly remarkably efficient, as well as constantly in demand. The sorry plight of xeroderma patients makes this abundantly clear. It may even be that the secret of old age lies in the quality of this crew, as suggested by the evidence of a striking correlation between life span and speed of DNA repair. As illustrated by the charts at the top of this page, such a correlation has been observed among different mammalian species, as well as within the order

These two relatives (top: house mouse, *Mus musculus*; bottom: the white-footed mouse, *Peromyscus leucopus*) differ strikingly in life span: *Mus* lives at most 3.5 years, *Peromyscus* as much as 8.3 years. It so happens that *Peromyscus* cells repair their DNA two and one-half times as fast as do those of *Mus*.

of primates (with humans the winners). It has even been found to exist between two kinds of mice.

To enjoy such excellent service is clearly an invaluable boon, but there are circumstances where it is tiresome. Cancer therapy is one. Many of the agents used in our fight against this disease—irradiation, intercalating drugs such as daunorubicin and Adriamycin, cross-linking drugs such as mustard gas and its derivatives—act by harming DNA and thereby blocking or derailing its transcription and replication. One of the reasons these efforts are so often thwarted lies in the ability of the target cells to correct the injuries inflicted on their DNA. And, of course, natural selection sees to it, in its utterly blind and insensitive way, that the cells best equipped to resist the therapy remain to spread and proliferate.

Another kind of service that cells maintain to protect the purity of their genome involves the recognition and destruction of foreign DNA. This is obviously very difficult if there is nothing special about the foreign DNA, and there are, in fact, innumerable occasions on which cells are misled into accepting foreign DNA brought in by viruses or other agents, sometimes to the point of yielding their whole genetic machinery to the invader. But some defense is possible, thanks to the restriction enzymes, so named because they restrict the compatibility between host and invader. Restriction enzymes are endodeoxyribonucleases that recognize and split certain specific sequences, usually of self-complementary palindromic structure. We have already encountered a similar kind of structure at the origin of replication. One of the first known restriction enzymes is EcoRI, extracted from the bacterium *E. coli*. It has the following specificity (arrows indicate site of splitting):

$$\downarrow$$
$$5'\text{--------}GAATTC\text{--------}3'$$
$$3'\text{--------}CTTAAG\text{--------}5'$$
$$\uparrow$$

Restriction enzymes have become of paramount practical importance because they allow DNA to be cut into well-defined segments, limited by the so-called restriction sites. These segments are used in DNA analysis and for genetic-engineering purposes. When, as in the case of EcoRI, the splitting is asymmetrical, the resulting segments have "sticky," or cohesive, ends. They tend to join by base-pairing with any other segment that has been produced by the same enzyme and, therefore, possesses the same ends. As we shall see, this property is widely exploited in genetic engineering. For these reasons, the discovery of the first restriction enzymes has spurred a massive hunt for similar enzymes with different specificities. Today, more than two hundred different restriction enzymes are known. All, so far, have been isolated from bacteria.

But what about their physiological role? Restriction sites are so simple that they are hardly likely to distinguish foreign from indigenous DNA. Rather, the opposite would be expected, since the invading DNA is usually

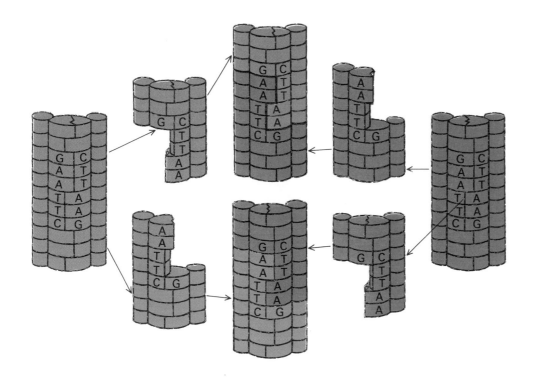

Hydrolysis of two different DNAs possessing the appropriate restriction site (GAATTC palindromic sequence) by the restriction enzyme EcoRI produces fragments with "sticky ends." Treatment with ligase reassociates the pieces at random, generating hybrids of the two DNAs, among other products. This technique is widely used in generic engineering (see Chapter 18). Note that, if the restriction site is methylated (on central adenines), endonuclease splitting by EcoRI is prevented. *E. coli* uses this device to protect its own DNA against the action of its EcoRI.

much smaller than the host-cell DNA and may therefore be expected to lack—and indeed often does lack—restriction sites that are present in the host DNA. The recognition trick, once again, depends on methylation. Bacteria that use a given restriction enzyme protect the corresponding restriction sites in their own genome by methylation. Restriction enzymes act only on unmethylated sequences. Of course, if the invader is clever enough—that is, if it has been retained by natural selection owing to that property—to methylate its own restriction sites, it will not be recognized as foreign. And so the battle goes on. Note that this is the third time we have encountered methylation as a protective device against nuclease action. Only moments ago, we met it as a means of avoiding excision of the correct sequence in a mismatch. In Chapter 16, we saw how it serves to guide the processing of rRNA. In Chapter 18, we will meet it in a different capacity: as a means of silencing DNA and of inhibiting its transcription.

18 | Recombination and Other Genetic Rearrangements

All that we have seen so far is geared to favor the strict preservation of genetic orthodoxy. In no state or church has information been policed with such rigor and deviation been prosecuted with as much severity as they are in the nucleus. Looked at from the point of view of the cell, such literalism makes sense. A living cell has a good thing going, and the dangers of change are so much greater than its possible benefits as to make heresy almost inevitably self-purging. Yet, diversity is the spice of life, variability the source of innovation in the biosphere. They are introduced into the genetic game by two kinds of accidents: some are truly accidental and fortuitous; others are made to happen by a genetically programmed enlistment of chance.

Shuffling Genetic Cards

It has been known since the beginning of this century that chromosomes may exchange segments. This happens during the meiotic maturation of germ cells, when, after duplication of their DNA content, paired chromatid dyads line up closely against each other and exchange homologous pieces by crossing-over before they separate and divide to form distinct haploid complements in the four daughter cells (Chapter 19). This phenomenon enormously enhances genetic diversification through sexual reproduction. Thanks to it, the haploid set carried by each spermatozoon or ovum does not simply consist of whole parental chromosomes selected at random from pairs of homologues, as would be the case

A mechanism for legitimate recombination.

1. A nick in a strand allows DNA polymerase to start elongating the 3′ end freed by the nick.

2. As the polymerase progresses, it displaces the nicked strand, as it does in a normal replication fork (Chapter 17).

3. The displaced strand becomes entangled with the complementary strand of the closely apposed allelic DNA, in turn displacing its counterpart as a loop.

4. At some stage the polymerase drops off, the displaced loop is cut off by endonuclease action, and the loose ends are stitched together by

ligase action (arrows). The two alleles are linked by a typical chiasma.

5. Rotation of the upper part of the chi-form is followed by the cutting of their connecting strands (arrows).

6. The cut strands are attached together in reciprocal fashion, completing the exchange of segments between the alleles.

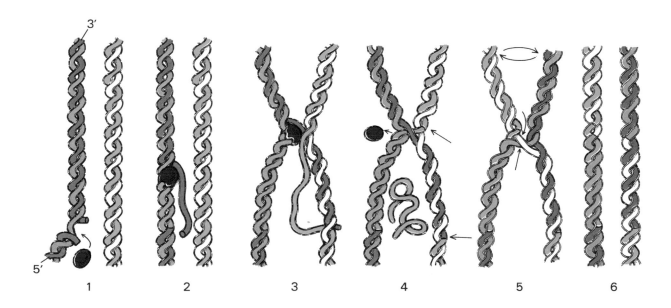

1   2   3   4   5   6

without crossing-over; but each individual chromosome is itself made of a number of fragments copied either from the father's or from the mother's contribution to the individual's genome.

A key characteristic of this recombination process, as it is called, is its faithfulness and precision. Barring accidents—the inevitable hazard of any natural process—exactly homologous pieces of the two chromosomes are exchanged. By looking at the way a DNA fiber is curled up and compressed into a stubby chromosomal rod, you can readily imagine that the exchange of segments between two adjacent chromosomes must be a complex process. But at least you can visualize its main molecular basis and, at the same time, understand why it is so extraordinarily precise. At some stage, preceded, as we shall see in Chapter 19, by parallel positioning of the two homologous chromosomes in close register with each other, some sort of hybridization must occur between a single-stranded piece of one chromosome and its complement in the other chromosome. Most likely, what causes the generation of a loose single strand is nicking followed by polymerase action, as happens in excision-repair (p. 332). Instead of being broken down, however, the displaced strand becomes entangled with the neighboring homologous chromosome. A possible mechanism leading to the formation of a chiasma and to the exchange of segments between the chromosomes is illustrated by the diagram shown above. This mechanism, it will be noted, depends entirely on familiar enzymes. Except for the initial embrace, which requires the participation of special recombination factors, subsequent events rely essentially on the same nuclease, topoisomerase, polymerase, and ligase activities as are involved in DNA replication and repair. More important, whichever the exact mechanism involved, it is clearly dependent on, and its precision guaranteed by, the close structural similarity

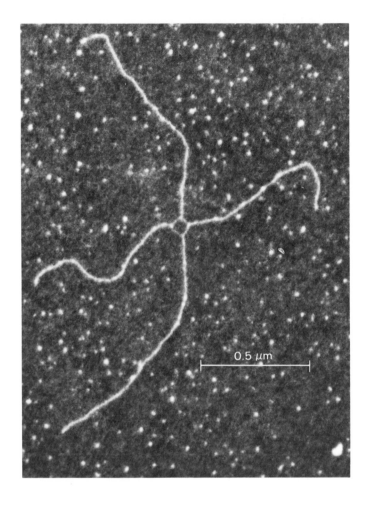

Chi-form created by two *E. coli* plasmids in the process of recombining. The central square consists of single-strand connections between homologous segments of the two plasmids (see the drawing on the preceding page). The connections are separated because the double-stranded DNA has become partially unwound in the cross-over region by the method of preparation.

0.5 µm

between the hybridizing DNA segments. All homologous germ-line chromosomes of a given species carry the same genes in the same order, with only occasional differences (mostly base-pair replacements) that distinguish two alleles from each other.

Bacteria, although they have only a single chromosome, can engage in similar recombination events whenever they receive a piece of homologous DNA by transformation, conjugation with another bacterium, or transduction (DNA transfer mediated by a virus). In most of these cases, recombination occurs by exchange of closely homologous segments, as in crossing-over. Events of this sort are now categorized under the general designation of "legitimate" recombination.

This term implies that "illegitimate" recombinations also take place. Suspected for some time, this fact has become one of the latest and most thrilling causes of excitement in a field that seems to have an almost inexhaustible store of surprises awaiting the explorers who dig deeper than did their predecessors. Apparently, certain genes or sets of genes can wander through the genome, accounting for an immense variety of evolutionary and developmental phenomena, probably including the production of cancer. In actual fact, the word illegitimate is not entirely deserved, for homology remains the basis of the recombination. But it is an internal homology, which concerns only the two ends of the wandering DNA segment. In all known cases, these two ends are identical, or nearly so, either as such (direct repeat) or in palindromic form (inverted repeat).

An example of a direct repeat would be:

$$5'\text{----ATCGCTC--------ATCGCTC----}3'$$
$$3'\text{----TAGCGAG--------TAGCGAG----}5'$$

And an inverted repeat:

$$5'\text{----ATCGCTC--------GAGCGAT----}3'$$
$$3'\text{----TAGCGAG--------CTCGCTA----}5'$$

This kind of homology may concern only a short sequence, as shown in the example above. But quite often it

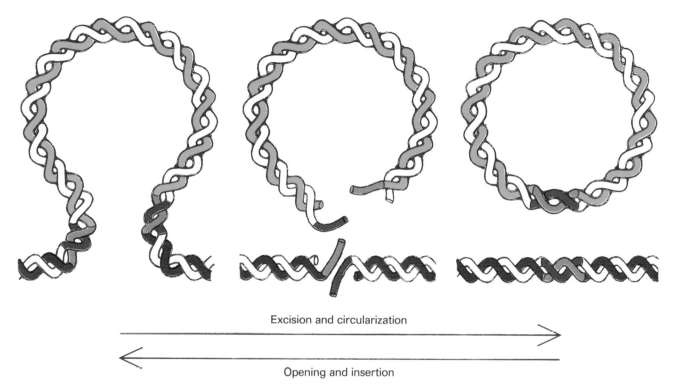

Excision and circularization

——————————————————————————→

←——————————————————————————

Opening and insertion

Conservative illegitimate recombination, with direct repeats, allows excision (with circularization) or insertion (with linearization) of a segment of DNA.

extends over much greater stretches, sometimes more than one thousand base pairs long.

Illegitimate recombinations are either conservative or duplicative. In the conservative type, the moving piece of DNA is translocated without duplication by what can be understood, at least in principle, in terms of an asymmetric nicking of the terminal repeats and their reclosing in a different way. If the repeat is direct, this mechanism will lead to excision and circularization of the piece of DNA flanked by the repeats or, conversely, to the insertion of such a circle into an homologous acceptor site. In essence, such a mechanism can account for the kind of "double life" led by certain viruses, either as independent, self-reproducing entities (circular viruses or phages) or as integrated parts of the host genome (proviruses or prophages). In practice, the excision and insertion mechanisms used by viruses are usually more complex.

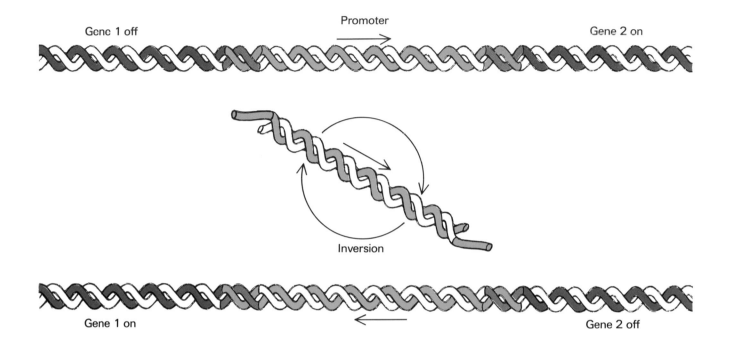

Gene 1 off   Promoter   Gene 2 on

Inversion

Gene 1 on   Gene 2 off

Conservative illegitimate recombination, with inverted repeats, allows inversion. A promoter on the inverton switches gene 1 or 2 on, depending on the position of the promoter.

With inverted repeats, the only possible conservative translocation is inversion. If the invertible segment, or inverton, bears an appropriately placed promoter on one of its strands, the position of this genetic "flip-flop" switch determines which of two genes that can be controlled by the promoter is to be transcribed. Such switches exist. Certain salmonellae, which are microbes that cause gastrointestinal infections, use this system to change from one to another flagellar protein. If the host becomes immune to the "flip" variant, switching over to the "flop" position will give new life to the infectious microorganism. We see here a simple example of antigenic variation. Trypanosomes, the protozoan agents of sleeping sickness, have developed this evasive defense trick to a fine art. They can change their surface coat protein many times in succession by switching on different genes. How they do this is not entirely clear but seems to involve du-

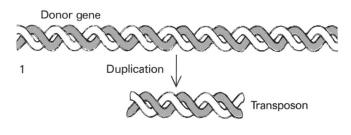

Donor gene

1   Duplication

                                    Transposon

Acceptor gene

2   Asymmetric nicking

Insertion of transposon

3   Gap filling creates direct repeats

Duplicative illegitimate recombination, or transposition.

1. A transposable segment of DNA is
copied by duplication.

2. An acceptor site is split asymmetri-
cally to receive the transposon.

3. Stitching requires gap-filling
replication of the single-stranded ends
of the acceptor site. The inserted
transposon thus becomes flanked by
direct repeats.

plicative transposition of genes from mute sites where
they are stored, but not transcribed, to expression sites.

As its name indicates, duplicative transposition involves making a copy of the movable DNA sequence and
inserting it into a new site while the original remains in
place. Sequences that are translocated in this way are
called transposons if they contain genes. The name insertion sequence designates shorter transposable sequences
not known to contain genes. However, the distinction
between the two is fuzzy. The term transposon will be
used for all transposable sequences that move duplica-
tively.

The mechanisms of duplicative transposition are complex and imperfectly understood. From what is known, it
appears that much of the information needed for transposition lies within the transposon itself, in particular in the
terminal repeats that exist in all known transposons. In
addition, a number of important transposons in the bacterial world include a gene that actually codes for
a special enzyme needed for transposition (transposase).
So far, nothing special has been detected in the insertion
site or target, except that transposition invariably leads to
duplication of a stretch from four to twelve base pairs
long in such a way that the inserted transposon is flanked
by direct repeats of this stretch. Asymmetric nicking of
the target, followed by replicative gap-filling, is the obvious explanation. The target sequences themselves are
highly variable, suggesting that transposons may enjoy
considerable freedom of movement.

Transposition accounts for a large number of genetic
exchanges mediated within and between bacteria by such
extrachromosomal, self-replicative entities as phages or
plasmids. In particular, as we shall see (p. 347), it plays a
major role in the spread of antibiotic resistance, since
many of the genes that code for enzymes that render their
owners resistant to some antibiotics are included in
transposable elements carried by plasmids.

In both prokaryotes and eukaryotes, transposition is
an important cause of evolutionary change. Transposons
can create all sorts of disturbances merely by their insertion in a certain site. They can cut in the middle of a gene
and so kill it, block or promote the transcription of neigh-

Genetic map of human globin genes. Hemoglobin, the oxygen-carrying protein of red blood cells, is a tetramer of general structure $\alpha_2\beta_2$. Each subunit resembles myoglobin (Chapter 2) and bears a heme group (Chapter 9). Early embryos, fetuses, and infants have different hemoglobins adapted to the manner in which they are supplied with oxygen. Different α and, especially, ß subunits, controlled by different genes, make up these different hemoglobins. The location of these genes on their bearer chromosomes corresponds to the order

in which they are expressed in the course of development. Note that all these genes (including the pseudogenes, or dead genes, $\psi\beta_2$, $\psi\beta_1$, and $\psi\alpha$) originate from a single ancestral gene that, it is estimated, underwent duplication some 500 million years ago. Since then, a large number of events—including duplications, transpositions, deletions, and other changes—took place to produce the complex situation shown on the map. Sequencing of the globins and their genes in different species have made it possible to time many of

these events in terms of the number of point mutations suffered by the genes since the event took place (see pp. 353-355). Such point mutations are responsible for additional diversification within a given species. Analysis of human hemoglobins has disclosed the existence of more than 270 variants, many of which are functionally abnormal. The most widespread of these abnormal hemoglobins is hemoglobin S, a ß-subunit variant responsible for sickle-cell anemia (see the electron micrograph on p. 352).

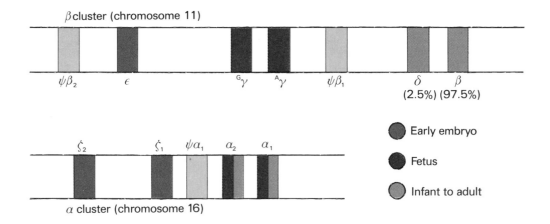

β cluster (chromosome 11)

$\psi\beta_2$ ϵ $^G\gamma$ $^A\gamma$ $\psi\beta_1$ δ β
(2.5%) (97.5%)

ζ_2 ζ_1 $\psi\alpha_1$ α_2 α_1

α cluster (chromosome 16)

● Early embryo

● Fetus

● Infant to adult

boring genes, or, alternatively, have the transcription of their own genes blocked or promoted by neighboring sequences. In addition, they have the unexplained ability of creating havoc around themselves in the form of deletions, duplications, or inversions of nearby DNA stretches.

In itself, the phenomenon of gene duplication that is an intrinsic part of all duplicative transpositions is of major evolutionary significance, because, once genes have been duplicated, their twin copies may subsequently evolve separately into two progressively different genes. The kinship between these genes can, however, be detected by

a comparison of their sequences or of those of their protein products. Examples of such evolutionary siblings abound. We encountered one with the two tubulins and another with calmodulin and troponin C.

An interesting example of gene duplication and evolution is provided by globin, the protein part of hemoglobin, the oxygen-carrying pigment of red blood cells. Globin is a tetrameric protein, made of two α and two β chains. The α and β genes are evolutionary offshoots of the same ancestral gene. But this is not all. Our genome contains several different α and β genes, which code for the different hemoglobins that are found at different stages of development (embryo, fetus, adult). These genes are situated on the same chromosome in the order in which they are expressed during development. Interspersed between them are a couple of "dead genes," or pseudogenes, siblings mutated to the point that they are no longer expressed. Finally, the evolutionary history of the globin genes has varied among different human subgroups, resulting in the appearance of numerous mutant hemoglobin molecules. These all conserve the ability to carry oxygen—otherwise, the mutation would, of course, be lethal—but some of them are functionally impaired. The best-known genetic disease caused by a disabling alteration of a globin gene is sickle-cell anemia, the result of a mutation widely distributed in people of African descent. The molecular basis of this mutation and the mechanism of its spreading will be examined later in this chapter (see p. 352).

The most remarkable known instance of gene shuffling occurs in the course of B-lymphocyte differentiation. As we saw in Chapter 3, this group of cells includes a very large number of distinct individuals, each of which recognizes a different antigen. Upon contact with a given antigen, the few cells that have a receptor for it on their surface are induced to multiply (the phenomenon of mitogenic stimulation was considered in Chapter 13) and give rise to a clone of identical plasma cells that manufacture an antibody directed against the inducing antigen.

As illustrated on the next page, antibodies are Y-shaped proteins made of two pairs of identical polypeptide chains, known as L (for light) and H (for heavy), joined by disulfide bonds. Outer parts of the branches of the Y, which include the N-terminal ends of both chains, make up the variable regions that determine the specificity of the antibody. The C-terminal parts of the two chains make up the constant region. Those of the heavy chains extend beyond those of the light chains to form the stem of the Y. The antibody repertoire of an organism encompasses an enormous number of different variable regions, each made, we must remember, by a distinct type of B lymphocyte. This number is believed to run into hundreds of millions, if not billions, to account for the fact that we can make antibodies against virtually any kind of antigen, natural or artificial, to which we are exposed. In contrast, the constant regions are the same for all antibodies in a given class, irrespective of their immunological specificity. But they vary in the H chain according to the class of antibodies (IgG, IgA, etc.). Certain H-chain constant regions are fitted with an extra C-terminal hydrophobic tail that allows anchoring of the immunoglobulin molecule in the plasma membrane to serve as mitogenic receptor.

Each antibody molecule needs at least 1 μm of DNA for its genetic encoding. Clearly, with its 2,000,000-odd μm of DNA, the genome of the stem cells (or of their embryonic progenitors, from which lymphocytes originate) cannot possibly accommodate the huge number of genes that code for antibodies. These genes must arise during differentiation of the lymphocytes, so that, starting from a relatively small common stock present in the stem cells, each mature cell ends up with a different gene. The new cell explorers who are now cutting their way through the dense thickets of eukaryotic nuclei with the help of restriction enzymes, cloning vehicles, and other genetic-engineering tools have identified this stock as consisting of seven distinct sets (three for the L chain and four for the H chain) of between half a dozen and a few hundred interchangeable parts, from which mature L and H genes are assembled by a special recombination process (see p. 344). A conservative estimate puts at 18 million the number of variants that can be generated by this mechanism. This number is further increased, perhaps as much as a thousandfold, by slight irregularities in the joining process and by mutations in certain highly vulnerable regions. Special splicing mechanisms at both the gene and the mRNA levels account for switching of classes in the H-chain constant region.

This example shows clearly that development is not simply a matter of turning certain genes on or off at the appropriate time and that the genome itself may undergo programmed rearrangements. A lymphocyte nucleus could definitely not be used for the kind of cloning experiment discussed in Chapter 16. We do not know to what extent gene maturation occurs in other cell types.

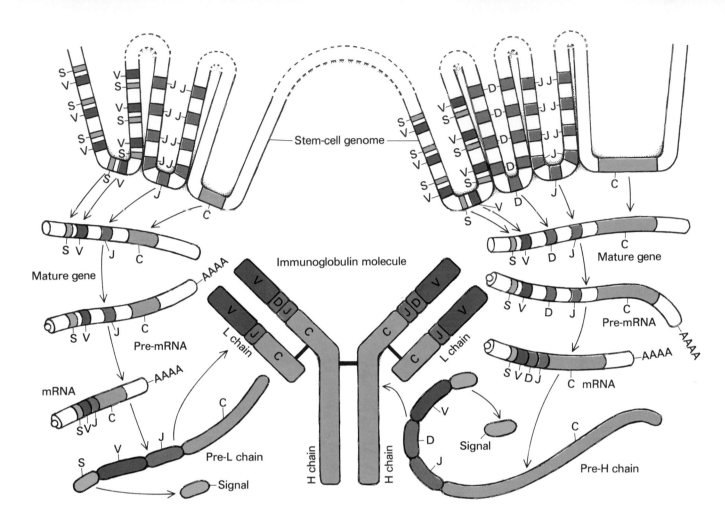

Immunoglobulin molecule

Stem-cell genome

Mature gene

Pre-mRNA

mRNA

Mature gene

Pre-mRNA

mRNA

Pre-L chain

Pre-H chain

L chain

L chain

H chain

H chain

Signal

Signal

Genetic differentiation during the maturation of B lymphocytes. An immunoglobulin molecule consists of two heavy (H) and two light (L) chains, linked together by disulfide bonds. These chains consist of an N-terminal variable part (V) and a C-terminal constant part (C), connected by a joining piece (J), with, in the H-chain only, a small additional piece D (for diversity). The stem-cell genome contains multiple variants of the L-chain V and J genes and of the H-chain V, J, and D genes. The V genes are preceded by a small S segment coding for an N-terminal signal piece (the two chains are secretory proteins made by ER-bound ribosomes). As lymphocytes mature, each differentiating cell constructs a particular L gene and a particular H gene of virtually unique structure by a special recombination process, dependent on strategically placed complementary sequences, that randomly selects one out of each set of gene segments and assembles them together with a C gene. Transcription of the gene produces a pre-mRNA that is spliced to give a mature mRNA, which is translated to the corresponding immunoglobulin chain with cotranslational removal of the signal piece (Chapter 15). The immunoglobulin molecule is then further put together by oxidative formation of disulfide bridges from cysteine residues, a process that takes place inside the ER lumen (Chapters 6 and 13). This illustration does not show how different C genes are used for the H chain to make different classes of antibodies. It is not known whether recombination occurs within a single chromosome, as illustrated, or by asymmetric exchange between two alleles during mitosis.

Ballast DNA

Should we stroll through a nucleus—assuming the terrain lent itself to such a desultory activity—and try, with the eagerness of neophyte scholars, to decode all its inscriptions, we would be astonished to find out how few of them make sense. Large chunks of DNA, made of millions of simple, repetitive sequences, completely fill certain regions, which correspond mostly to the centromeric parts of the chromosomes (see Chapter 19). Known as constitutive heterochromatin, they are permanently condensed and almost certainly instruct no process other than their own replication. Then there are the long, untranscribed linkers between the genes; the dead genes; the introns, which often greatly exceed the exons in length; the untranslated leader and tail sequences of the mRNAs; the large pieces that are excised in the maturation of ribosomal and other functional RNAs. Repeats also occur in those unexpressed pieces of DNA that are interspersed between coding parts. Best known is the *Alu* family, a characteristic of human DNA, so named because its members can be neatly carved out of the genome by means of a restriction enzyme called Alu I. It includes some 300,000 to 500,000 identical, or nearly identical, sequences of about 300 base pairs. Other repeats, found in fewer numbers, are of the type that flank transposons.

Taken together, all the DNA in the nucleus that is not overtly expressed as RNA or protein may amount to more than 90 per cent of the total. The very size of the genome makes this clear. With a haploid content of almost 3 billion base pairs, we have enough DNA in our nuclei to code for 1 million different proteins of about 1,000 amino-acid residues. In fact—not counting the immunoglobulins, which represent an exceptional case, as we have just seen—we manufacture less than one-tenth this number throughout our existence, from conception to death. From which it follows that more than 90 per cent of the DNA that we diligently replicate at each cell division never becomes translated into an actual mediator of phenotypic manifestation. The puzzle is compounded by the fact that closely related species may have very different DNA contents. This is called the C-value paradox.

(The term C-value goes back to the days when it was first found that different cells in a given species have the same—constant—amount of DNA in their nuclei.)

The interpretation we may put on these intriguing facts depends on what particular Darwinian sect we belong to. We may, with the "strict selectionists," put our faith in natural selection and refuse to believe that the nucleus would keep perpetuating a lot of rubbish with no selective advantage to the species. Or we could join the "genetic drifters" and argue that natural selection has no power over what is simply useless; it eliminates only the harmful. And so functionless DNA could, whatever its origin, be carried along passively and just drift, for a very long time, before turning into more than an insignificant inconvenience.

Truth probably lies somewhere in between these two extreme views. There is no doubt that DNA contains a great deal of information that, although not expressed in a transcription or translation product, nevertheless plays an essential role in the operation and regulation of nuclear activities—in the form of replication origins, promoters, and terminators; of binding sites for enzymes, repressors, or other regulatory agents; and of many other, as yet undeciphered, code words that are used in the intricate, four-dimensional communication networks of living cells and organisms. Many of these signals may be expected, like road signs, to fall into a limited number of characteristic classes and to be interspersed throughout the genome, thus accounting for at least part of the repeats. The *Alu* family, for example, is believed by some to contain replication origins, which must, indeed, number in the hundreds of thousands in a human nucleus. The highly clustered repeats of constitutive heterochromatin obviously do not function as road signs, but they could still play an important role in chromosome organization. As we shall see in Chapter 19, the centromeric region with which constitutive heterochromatin is preferentially associated may well sleep during the whole of interphase, but it wakes up to become a star performer when the mitotic drama is enacted.

Similarly, there must be a variety of regulatory signals among the DNA sequences that are transcribed but do not end up as part of a functional RNA or protein. Specific signs are needed to attract and direct the numerous factors that are involved in RNA processing and translocation. Others serve to guide the finished RNAs to their terminal destination in or on ribosomes, or elsewhere, and to instruct the systems with which they interact. Here, again, we may expect to find repeats—for instance, at the 5' ends of mRNAs. Down at the polypeptide level, finally, we have seen that those sequences that are discarded in the fashioning of a protein are often removed only after they have carried out an important function, such as signaling (Chapter 15).

Altogether, these regulatory sequences may well add up to several times the size of the overtly expressed part of the genome. But it is not likely that they account for all the noncoding DNA. A frequently invoked criterion of functional usefulness is the degree of evolutionary conservation. Sequences that have not changed much in the course of evolution, as indicated by close similarities between distant species, are seen as important. On the other hand, those that have changed rapidly are considered not to be meaningful. Many introns fall in the second category.

As often happens when facts are scarce and beliefs strong, hardliners of either one or the other creed refuse to take evolutionary conservation as an absolute criterion, each for different reasons. The strict selectionists point out that meaningless is not necessarily synonymous with functionless. Noncoding DNA could play a role by its mere bulk, as ballast. As pointed out in Chapter 16, the flexibility conferred on the eukaryotic genome by the existence of long intervening sequences may well have been essential in the evolution of plants and animals. The drifters, on the other hand, will tell you that conservation does not automatically imply usefulness; that it is in the nature of genes to ensure their own replication, and even multiplication thanks to transposition; and that natural selection has no control over these phenomena as long as they do not affect a phenotype's ability to reproduce the genotype. This view has become known as the "selfish gene" or "parasitic gene" concept. Dead genes, of which several have already been recognized (in globin, for instance, as mentioned on p. 342), are typical examples of selfish genes.

As tourists, you are not expected to take a stand on these matters. Nor does your guide command enough expert authority to advise you. It would seem, however, that, until more facts are known, we would be wise to take the ecumenical position that there is probably some truth in each point of view.

Viruses, Phages, Plasmids, and Other Flying Genes

A virus may be defined as a piece of genetic material, either DNA or RNA, surrounded by a protein coat, or capsid (Greek *kaps*, box), and, in some cases, by a membranous envelope. The main function of these wrappings is to mediate the penetration of the genetic material into some recipient cell in which multiplication of the virus will occur. If the recipient cell is a prokaryote, the virus is often called phage. This term is derived from bacteriophage or "bacterium eater," the word originally proposed by the French bacteriologist Félix d'Hérelle, who discovered these entities through their ability to kill bacteria. A plasmid is a small piece of circular DNA that is not integrated in a chromosome and can be exchanged between cells. Plasmids are found mostly in bacteria, but they are known to be present also in eukaryotes, as, for instance, in yeast.

Long known as agents of major diseases—among them smallpox, measles, poliomyelitis, hepatitis, influenza, and rabies—viruses and their counterparts in the prokaryotic world, the bacteriophages, have become the darlings of molecular biologists. Thanks to their relative simplicity (the term "relative" is appropriate, because even the simplest of viruses is highly complex), they have served in an invaluable manner to initiate cell explorers into the mysteries of the gene. They make up an extraordinarily diversified group, which our other commitments allow us to consider only in the most general of terms.

Viruses travel light, according to a strictly economical policy. They rely almost entirely on the enzymic machinery and metabolite supply of their host cells and carry in their genetic baggage only the blueprints of their capsid and envelope proteins, together with those of such enzymes and other protein factors they may need to subvert their hosts and to ensure their own replication. The smaller phages and viruses require only half a dozen genes (or fewer) for this purpose, whereas the larger ones may code for more than a hundred enzymes, including a complete DNA replication kit and, sometimes, a rather fiendish system that radically shuts off DNA replication by the host and leaves the virus in full control. All these proteins are synthesized by the host's protein-manufacturing machinery from transcripts of the viral genome.

Viruses have their genetic information encoded in either DNA or RNA. Most viral DNAs are double stranded and are replicated and transcribed by different variants of the general mechanisms described in Chapters 16 and 17. A few small DNA phages are single stranded. The first thing that happens when such a phage enters a cell is the formation of a complementary strand. The resulting duplex, or replicative form, serves in transcription and in the production of new viral strands.

DNA viruses may mount two types of attacks. In the virulent kind, the host cell is completely taken over by the virus and used for viral multiplication until it can no longer survive. The infected cell eventually disintegrates (lysis), and the viral progeny are released to infect other cells. In the nonvirulent, or temperate, kind, the viral DNA becomes integrated in the host cell's genome by some recombination or transposition process and replicates with it. This occurs in such a way that the genes that control lysis are shut off. Such integration may have a number of fateful consequences. It may serve to spread a latent infection (lysogeny) in a whole population of cells without a detectable symptom until some stress, such as exposure to ultraviolet light, causes the viral DNA to be excised from the genome and to initiate a catastrophic virulent infection. Another possible effect, if the viral DNA has the properties of a transposon and can wander through the genome, is multiple mutations. In animal cells, DNA viruses that become integrated in the genome tend to cause cancerous transformations. A particularly simple example of such an oncogenic virus is SV40, already encountered in Chapter 16. This virus is normally found in certain Asian monkeys (SV stands for simian virus), in which it grows without causing any apparent illness. In other species, or in cultured cells, SV40 may, according to circumstances, cause a lethal infection or induce a cancerous tumor. It develops inside the nucleus of infected cells, where it behaves very much like a small, local chromosome. In particular, it combines with histones to form nucleosomes, and its transcription products are spliced, capped, and polyadenylated, as are most eukaryotic mRNAs. We saw in Chapter 16 how different modes of splicing allow five distinct proteins to be made from only two transcripts of the SV40 DNA. Of them, the VP proteins are viral proteins, and the t and T proteins mediate the cancerous transformation.

Plasmids differ from viruses by lacking a capsid and being largely uninfective for this reason. But they can be exchanged between bacterial cells during conjugation and may be made to enter cells under special conditions, which are now widely exploited in genetic engineering and may well obtain sometimes in nature. In any case, there is no doubt that plasmids move around a great deal. They are fairly well behaved in the cells that they invade and generally replicate to only a few copies. Nevertheless, they can cause effects of very great importance. One reason is that a number of proteins that mediate resistance to antibiotics, as well as certain highly poisonous toxins produced by bacteria, are encoded by plasmid-associated genes. Thus plasmids may confer redoubtable properties on pathogenic microbes. Another property of plasmids, and of some viruses, is that they can serve as vectors for "passenger" or "hitchhiking" genes, which are removed from the genome of one cell and transferred to that of another by transposition phenomena (transduction).

With rare exceptions (an example is reovirus), RNA viruses are single stranded. They all have to code for a special polymerase, since cells do not possess enzymes capable of copying an RNA template. In the usual virulent viruses, this polymerase makes a complementary RNA strand, starting with a single nucleoside triphosphate as primer. It resembles transcriptases and primases, therefore, with the difference that it is instructed by an RNA strand instead of by a DNA strand. In one class, of

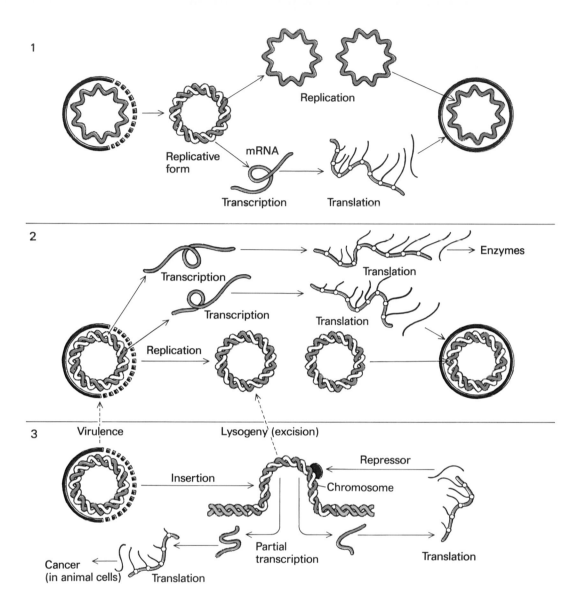

1

Replication

Replicative form

mRNA

Transcription   Translation

2

Transcription   Translation → Enzymes

Transcription   Translation

Replication

3   Virulence   Lysogeny (excision)

Insertion   Repressor

Chromosome

Partial transcription

Cancer (in animal cells)   Translation   Translation

Examples of viral growth cycles: DNA viruses.

1. Single-stranded (e.g., phage φX174). The viral DNA is replicated to a double-stranded replicative form, which serves as template for the synthesis of new viral DNA and for transcription to mRNAs that are translated into viral proteins.

2. Double-stranded, virulent (e.g., phage T₄, adenovirus). The viral DNA is replicated and transcribed. The mRNAs are translated into viral proteins, as well as enzymes used by the virus to establish dominance over the host cell.

3. Double-stranded, temperate (e.g., phage λ, simian virus SV40). The viral DNA is inserted into a chromosome. In this situation, transcription of the DNA is restricted by a repressor encoded by the virus, and replication of the viral DNA occurs only in concert with duplication of the chromosome. There is no proliferation of the virus. This kind of behavior is observed only in certain host cells or under certain

conditions. In other circumstances, the virus is virulent (as in part 2). A temperate virus may be returned to the virulent state (lysogeny) by treatments that lead to excision of the viral DNA. With animal viruses of this type, insertion of the viral DNA into a chromosome may cause the development of cancer.

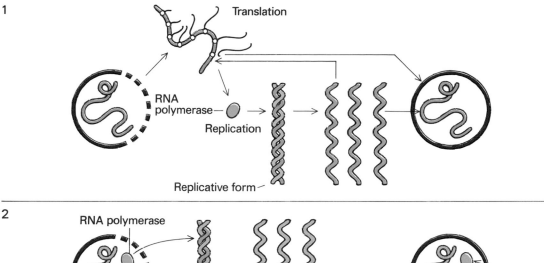

1

Translation

RNA polymerase

Replication

Replicative form

2

RNA polymerase

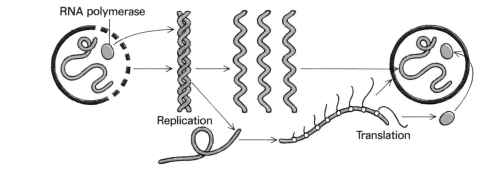

Replication

Translation

3

Reverse transcriptase

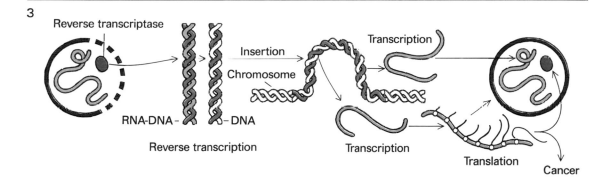

Insertion

Chromosome

Transcription

RNA-DNA

DNA

Reverse transcription

Transcription

Translation

Cancer

Examples of viral growth cycles: RNA viruses.

1. Single-stranded, plus (e.g., phage f₂, tobacco mosaic virus, poliomyelitis virus). The viral RNA is an mRNA (including cap and poly-A tail in eukaryotic viruses). Its translation produces viral proteins and the RNA polymerase needed for replication of the viral RNA by way of a double-stranded replicative form.

2. Single-stranded, minus (e.g., measles, influenza, rabies viruses). The viral RNA does not serve as mRNA and must first be replicated by a virus-associated polymerase. The plus strands that are generated are translated into viral proteins, and the double-stranded replicative form serves for the synthesis of new viral minus strands.

3. Single-stranded retroviruses (e.g., Rous sarcoma virus). The viral RNA is

of the minus type and is transcribed by a virus-associated reverse transcriptase to give a hybrid RNA-DNA duplex, which is further replicated into double-stranded DNA, which in turn is inserted into a chromosome. New virus particles are then formed by transcription and translation of appropriate parts of the incorporated viral genome. If the virus carries an oncogene, its insertion causes a highly virulent cancer. If not, cancer

may still ensue in rare cases as a result of the activation of an indigenous proto-oncogene by the inserted virus.

4. Double-stranded (e.g., reovirus). Not shown, this type of virus behaves essentially like a double-stranded replicative form (as in part 2), which is replicated and transcribed with the help of a virus-associated RNA polymerase.

which polio virus is a member, the viral RNA itself serves as mRNA. It is translated soon after infection to provide, among other proteins, the RNA polymerase needed for its replication. In the other class of single-stranded viruses, exemplified by rabies, the messages for the viral proteins, including the polymerase, are encoded in the strand complementary to the viral RNA. Such viruses need a preformed polymerase to make the complementary mRNA that will induce the cell to make more polymerase molecules. They carry this enzyme with them in their capsid luggage. Both virus types need a double-stranded intermediate (replicative form) for replication, as do the single-stranded DNA viruses. Double-stranded RNA viruses also carry their polymerase with them, to produce mRNA molecules from one of the two strands.

There is a special group of animal-cell RNA viruses, known as retroviruses or sometimes as oncorna viruses (from <u>onco</u>genic <u>RNA</u>), in which the virus-coded polymerase is a DNA polymerase. It is called reverse transcriptase, because it makes complementary DNA (cDNA) from an RNA template, in contrast to transcriptase, which makes RNA from a DNA template. The single-stranded DNA transcribed from the viral RNA is replicated to form a duplex and is then integrated into the genome by some kind of transposition process. This process is essential for the multiplication of the virus, which takes place by normal transcription from the integrated cDNA. Occasionally, integration of the cDNA in the genome results in the cancerous transformation of the infected cell by a mechanism that is currently raising an enormous amount of interest and may be related to gene transposition (see pp. 365–367).

Playing Evolutionary Roulette

The game has been on for more than 4 billion years. Barring a cosmic cataclysm, it is due to continue for at least as long. It is being played simultaneously over the whole surface of our planet by myriads of living forms ceaselessly engaged in replicating, transcribing, and translating genes, as well as recombining them in various

ways. There are so many twists to the game, so many possible moves, as to put it forever beyond the grasp of any conceivable brain, whether natural or artificial. Yet the rules are extremely simple and can be summed up in three words: fidelity, variability, selection.

Fidelity, we have seen, is a compelling condition, especially as regards DNA replication. No species can survive if its genome is not handed over essentially intact from generation to generation. Understandably, such species as do survive have available elaborate copying, proofreading, and other controlling devices that ensure the required level of fidelity. It is remarkable that these devices rely almost exclusively on the correct fitting of two pairs of small planar molecules joined at their edges by two or three weak electrostatic bonds. No chemist in his right senses would trust such a flimsy arrangement. Proteins, however, are endowed with powers of discrimination undreamed of by any chemist. But even proteins can err and, furthermore, no safeguard exists that provides absolute protection against every possible sort of accident.

Mistakes and accidents are the sources of variability, which is just as important as fidelity, provided it is kept within acceptable bounds. There are two kinds of deaths, one caused by slavishness, the other by laxity. The game of life is played in the narrow strait that separates the two. It is there that have been generated all the changes whereby living beings have diversified, adjusted to different ecological niches, adapted to new conditions, evolved, and progressed.

Genes are the exclusive targets of these changes, with, as the only possible exception, certain multimolecular assemblies that act as templates for homomorphic growth. The pattern of implantation of cilia on the surface of certain protozoa belongs to this rare category. And so, perhaps, does the assembly of membranes, although this requires further validation. On the whole, however, hereditary changes are, with overwhelming frequency, alterations of genes: mutations. They have as a fundamental characteristic that they occur by chance.

Coming at the end of this tour, such an assertion cannot but sound totally incredible. How many times did we stop in breathless admiration before some piece of molecular machinery that could be described only as "exquisitely designed" or some sort of equivalent? And now we are asked to see in all those wonders nothing but the products of blind, fortuitous events. It is simply ludicrous. Yet the proof is in front of our eyes, both in the actual manner in which genes are seen to operate and in the historical record that this operation has left in their structure and in that of their products.

What converts randomness into order is natural selection: any genetic change that enhances the survival potential of the individual concerned—more specifically, its ability to produce progeny—will tend to be retained at the expense of those that do not. This statement is, you will notice, essentially tautological and requires no demonstration. And it does not seem to matter that favorable changes happen only very infrequently compared with the others. The game of life is played at high speed, with huge numbers, over indefinite time. It can afford to wait for the occasional stroke of luck. And so it is, to quote the beautiful words of Jacques Monod (see p. 357), that "out of a source of noise, selection has been able alone to extract all the music of the biosphere."

This noise is of two kinds, depending on whether a single base or a whole stretch of DNA is affected. Changes of the first kind are called point mutations. It is easy to see by looking at the genetic code (Chapter 15) what the possible consequences may be when one base is replaced by another in a genetic message. Four types may be distinguished:

1. The altered codon is replaced by a synonym, as CAU by CAC, which likewise calls for histidine. The mutation is silent. It will appear only in the DNA and in its transcript but not in its protein translation product.

2. The altered codon specifies a new amino acid, but the biological properties of the modified protein remain unaffected. For example, a change from CUU to AUU, which replaces leucine by the closely similar isoleucine, is most unlikely to have a significant effect on the structure or function of the protein concerned. Mutations of this sort are called neutral, or indifferent.

We discovered in Chapter 15 that the structure of the genetic code is such as to favor mutational indifference, a strong argument in support of the theory that the code is itself a product of natural selection.

3. The altered codon specifies a new amino acid, but in this case the biological properties of the affected protein are significantly modified, usually for the worse. Mutations that change the electrical charge of a residue or greatly modify its hydrophilic or hydrophobic character are likely to have such effects. For instance, a change from GAG to GUG will substitute valine (uncharged, hydrophobic) for glutamic acid (negatively charged, hydrophilic). Such a change suffices to convert the normal human hemoglobin A into the severely pathological hemoglobin S, which is characteristic of sickle-cell anemia.

4. The altered codon is converted into a stop sign (e.g., UAC into UAG) resulting in abortive translation of the message, with the production of a usually inactive N-terminal fragment of the original polypeptide.

The consequences of mutations of the third and fourth kinds depend on how they affect the survival and reproductive potential of their victims. A priori, one would expect them to be lethal, or at least deleterious. But this is not necessarily so. It is remarkable that mammals lack many enzymes that are present in the lowly bacteria. Humans even lack some enzymes that are found in most of their mammalian relatives and, as a consequence, are subject to serious additional liabilities. We are, with other primates and, strangely enough, the Dalmatian dog, the only potential gout-sufferers among the mammals, because we lost at evolutionary roulette the capacity to break down uric acid. And we share only with our primate cousins and with guinea pigs the inability to manufacture vitamin C and the resulting distinction of being subject to scurvy. Clearly, these handicaps have not prevented us from getting on in the world. Perhaps—who knows?—they may even have helped, by conferring a selective premium on some genetically endowed piece of resourcefulness needed to overcome the handicaps.

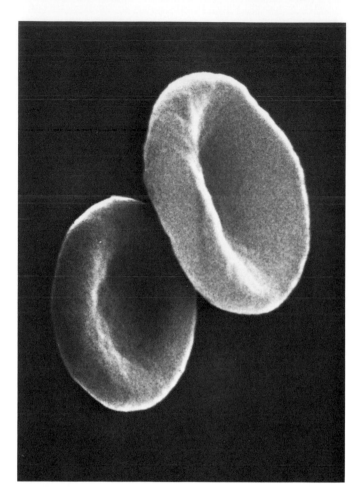

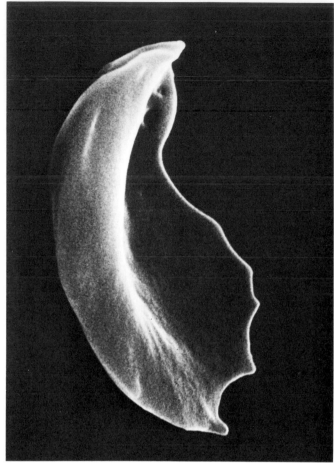

Sickle-cell anemia, the prototype of a "molecular disease." The scanning electron micrograph at the left shows the biconcave disk structure of normal red blood cells. That on the right illustrates "sickling," the structural deformation that the cells undergo at low oxygen tension in bearers of the disease. Such cells are abnormally fragile and suffer rapid destruction. Severe anemia and various vascular complications, leading to early death, are the final result. Remarkably, this dramatic unfolding of events is the consequence of a single amino-acid substitution (valine replaces glutamic acid in position 6 of the ß chain) in hemoglobin. This abnormal protein can still function as oxygen carrier but tends to precipitate out of solution when it has released its oxygen. This precipitation is responsible for the alteration in cell shape. The sickling trait is widespread in the black population. Its spreading is believed to have been favored by the relative resistance of heterozygotes to malaria.

Another important point is sexual reproduction and the resulting diploidy of the offspring. Many defective genes are deleterious or lethal only in homozygotes, a fact that is not likely to oppose their spreading by phenotypically normal heterozygote carriers until these carriers start making up a significant proportion of an interbreeding population. Sometimes the defective gene may even confer an advantage on the heterozygote. It is believed that the high frequency of the sickle-cell gene in people of African descent is due to the protective effect that it exerts against malaria in the heterozygotes.

Point mutations are largely responsible for what is sometimes called micro- or molecular evolution—that is, the progressive replacement, in the course of evolutionary time, of bases in homologous genes and of amino acids in the corresponding proteins. But they probably had little to

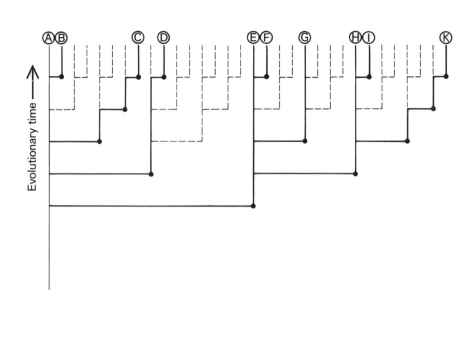

A	0									
B	1	0								
C	3	4	0							
D	2	3	5	0						
E	1	2	4	3	0					
F	2	3	5	4	1	0				
G	2	3	5	4	1	2	0			
H	2	3	5	4	1	2	2	0		
I	3	4	6	5	2	3	3	1	0	
K	5	6	8	7	4	5	5	3	4	0
	A	B	C	D	E	F	G	H	I	K

Number of amino-acid substitutions

Principle of molecular dating. The "tree" at the top illustrates a brief part in the evolutionary history of a hypothetical protein molecule. Each ramification point in the tree is the result of a mutation leading to the replacement of one amino acid by another. The mutated branch is indicated by a dot. Proteins A through K are evolutionary siblings differing from each other by the number of amino-acid substitutions indicated in the table below. The dotted lines represent extinct proteins. The sequencing of homologous proteins in different species yields the kind of information given in the table. The most probable phylogenetic tree can then be constructed from the data. The method

has many complications arising from the structure of the genetic code, the possibility of more than one mutation affecting the same site, the occurrence of mutations other than simple amino-acid substitutions (deletions, insertions, inversions), and the fact that different trees can accommodate the same data. But, as more and more proteins are being sequenced, the information assembled is becoming increasingly solid and secure. Such investigations are now being extended, with similar results, to genes and to RNAs. Soon, a highly detailed history of living organisms on earth will be available, as deciphered from the structure of their macromolecules.

do with the broader evolutionary phenomenon that has led to the development of increasingly complex forms of life. They seem to have happened with the same frequency throughout evolution and to bear no correlation to any of the events that have, at certain times, led rather abruptly to the appearance of new species.

This is so true that sequencing has become a favorite tool of evolutionary genealogists. Simply by counting the number of differences—bases in a nucleic acid or amino acids in a protein—between two homologous molecules belonging to different species and applying a suitable correction to account for silent mutations, consecutive replacements at the same site, and other possible complications, this new breed of historians will tell you how long ago the two species diverged from a common ancestor. The results of this elegant molecular-dating procedure

Comparative biochemistry of cyto-chromes. The table below gives the number of amino-acid substitutions in cytochrome c from twenty-five different species, including mammals and other vertebrates, invertebrates, plants, and fungi. The computer-generated phylogenetic tree of cytochrome c shown at the right is derived from sequencing data (most of which appear in the table below). In this calculation, amino-acid substitutions have been converted into the minimum number of nucleotide substitutions in DNA (according to the genetic code) needed to account for the observed amino-acid replacements. The branches are measured in terms of these substitutions. Agreement between molecular and more conventional methods of dating based on the fossil record is not perfect but remarkably close in view of the fact that only a single protein is considered.

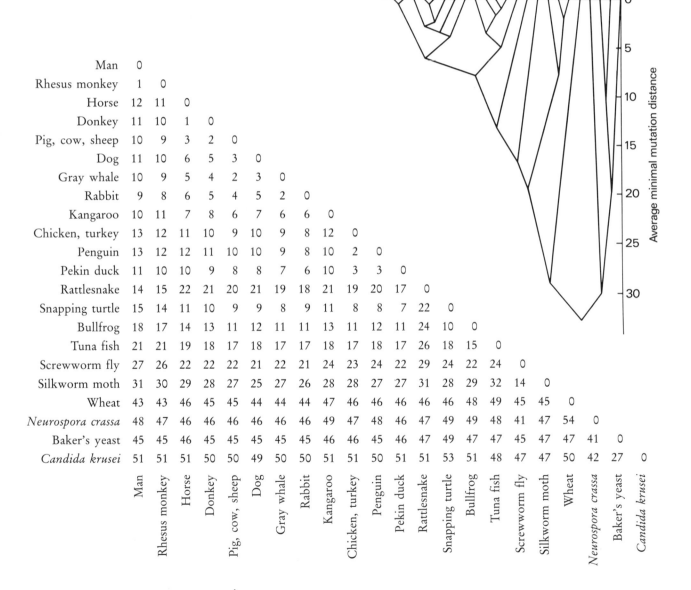

	Man	Rhesus monkey	Horse	Donkey	Pig, cow, sheep	Dog	Gray whale	Rabbit	Kangaroo	Chicken, turkey	Penguin	Pekin duck	Rattlesnake	Snapping turtle	Bullfrog	Tuna fish	Screwworm fly	Silkworm moth	Wheat	Neurospora crassa	Baker's yeast	Candida krusei
Man	0																					
Rhesus monkey	1	0																				
Horse	12	11	0																			
Donkey	11	10	1	0																		
Pig, cow, sheep	10	9	3	2	0																	
Dog	11	10	6	5	3	0																
Gray whale	10	9	5	4	2	3	0															
Rabbit	9	8	6	5	4	5	2	0														
Kangaroo	10	11	7	8	6	7	6	6	0													
Chicken, turkey	13	12	11	10	9	10	9	8	12	0												
Penguin	13	12	12	11	10	10	9	8	10	2	0											
Pekin duck	11	10	10	9	8	8	7	6	10	3	3	0										
Rattlesnake	14	15	22	21	20	21	19	18	21	19	20	17	0									
Snapping turtle	15	14	11	10	9	9	8	9	11	8	8	7	22	0								
Bullfrog	18	17	14	13	11	12	11	11	13	11	12	11	24	10	0							
Tuna fish	21	21	19	18	17	18	17	17	18	17	18	17	26	18	15	0						
Screwworm fly	27	26	22	22	22	21	22	21	24	23	24	22	29	24	22	24	0					
Silkworm moth	31	30	29	28	27	25	27	26	28	28	27	27	31	28	29	32	14	0				
Wheat	43	43	46	45	45	44	44	44	47	46	46	46	46	46	48	49	45	45	0			
Neurospora crassa	48	47	46	46	46	46	46	46	49	47	48	46	47	49	49	48	41	47	54	0		
Baker's yeast	45	45	46	45	45	45	45	45	46	46	45	46	47	49	47	47	45	47	47	41	0	
Candida krusei	51	51	51	50	50	49	50	50	51	51	50	51	51	53	51	48	47	47	50	42	27	0

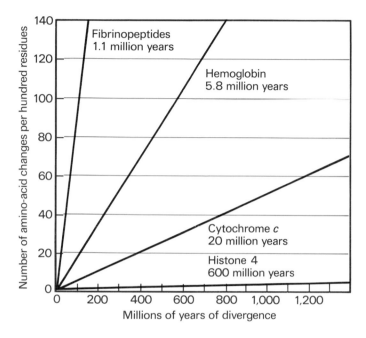

The linear relationship between molecular and fossil clocks of evolution suggests that, on an average and in first approximation, the mutation rate has remained constant throughout evolution and is the same in different species. More refined analysis indicates that this is not strictly true. What is striking in the graph, however, is the wide difference in the slopes of the different lines. This difference is not primarily an expression of differences in the mutability of the corresponding genes (although such differences may exist) but reveals mostly the degree to which changes in the amino-acid sequences of the proteins can be tolerated without loss of functional properties (nonviable mutants).

agree remarkably well with those of conventional paleontology. Not being restricted to the few traits that have been preserved in the fossil record, they have allowed us to extend, complete, and sometimes even revise the classical phylogenetic reconstructions. The relative time scale does, however, vary, sometimes considerably, from one molecular variety to another. Extremes among the proteins are the polypeptides (fibrinopeptides) that make up the meshwork of blood clots, which have suffered one amino-acid substitution in one hundred residues every 1.1 million years, and histone H4, which has needed about 600 million years for a single such replacement to occur. In between are hemoglobin, with 5.8 million years, and cytochrome c, with 20 million years.

These differences do not primarily reflect differences in the mutability of the genes concerned, although such differences may exist; rather, they reveal the stringency of the structural constraints on the functional properties of the corresponding proteins. Most likely, the genes coding for these proteins all mutated with about the same frequency, which sufficed, judging from the fibrinopeptides, to bring about at least one amino-acid substitution per hundred residues in their respective products every million years. However, for every hundred such substitutions that could be accommodated by the fibrinopeptides without loss of function (indifferent mutations), some eighty proved severely disabling to hemoglobin, ninety-five to cytochrome c, and essentially all to histone H4. They were consequently eliminated by natural selection, leaving no trace in contemporary living organisms.

Fascinating as they are, these examples illustrate only the negative, conservative aspect of natural selection—its strong tendency to keep things as they are and to allow only changes that make no difference. Clearly, such timorous toe-dipping into the waters of the unknown as is afforded by point mutations cannot account for the leaps that have propelled the upward surge of evolution. Something more drastic must have been at work.

That something is presumably represented by the deletions, the duplications, the recombinations, the transpositions, the splicings, the exchanges, the viral invasions, and all the other genetic migratory events that recent explorations have brought to light. Most of the time, such imprudent tamperings with orthodoxy must have been sanctioned without delay. But on certain rare occasions, perhaps coincidental with a sudden climatic or other environmental change, the freakish misfit proved to be better adapted than its "normal" congeners. It became a phagocytic hunter in a nutrient-depleted sea, or a fish out of water, or a tree dweller exiled to the savanna. Sometimes the resulting adaptations had far-reaching consequences.

All these events are buried in the depths of geological ages, and we will probably never get to know exactly how they happened, although we should not, perhaps, underestimate the power of molecular-biological analysis in reconstructing the past. Even more difficult to unearth are the events that preceded, and led to, the appearance of the first living cells. This difficulty has, however, not deterred biomolecular archaeologists from digging for clues. Already they have made the fundamental observation that many of the small organic molecules that are found in living matter, including amino acids, sugars, purines, and pyrimidines, can arise spontaneously under conditions mimicking those believed to have prevailed on the earth's surface in prebiotic times. Even primitive polymers have been generated in this way, among them oligonucleotides of sorts, as well as "proteinoids" endowed with catalytic activities resembling those of enzymes.

Mind you, this has nothing to do with creating life in the test tube, or even reproducing anything remotely resembling the mixture of molecules one gets from breaking apart a living organism. These artificial mixtures are really very "dirty." In addition to substances authentically found in living beings, they contain all kinds of other molecules, assembled in a very haphazard way. They could, however, approximate to some extent the "primeval soup" out of which the first living cells arose. How this emergence took place is a matter of conjecture, but it most likely involved, on a simpler chemical level, the same cardinal rules of fidelity, variability, and selection that governed biological evolution. Primitive self-maintaining and self-correcting systems must have formed and evolved progressively into dynamic structures of increasing complexity and stability. Other possibilities have been

considered, including, as noted in Chapter 10, insemination of the earth by germs from outer space (panspermia). But this only pushes the problem back to the origin of those germs and stretches the available duration hardly more than three of four times, unless you reject the "big bang" theory and believe, with the iconoclastic British astronomer Fred Hoyle, in a steady-state universe of "enormous antiquity," whatever is meant by that expression.

In actual fact, the age of the universe is essentially irrelevant to the discussion. If you equate the probability of the birth of a bacterial cell to that of the chance assembly of its component atoms, even eternity will not suffice to produce one for you. So you might as well accept, as do most scientists, that the process was completed in no more than 1 billion years and that it took place entirely on the surface of our planet, to produce, as early as 3.3 billion years ago, the bacteriumlike organisms revealed by fossil traces. After that, bacteria ruled the world alone for more than 2 billion years, during which they developed such important attributes as photosynthesis, oxidative phosphorylation, and, presumably, the phagocytic way of life; at the same time they profoundly changed the properties of the atmosphere and of the earth's surface. And then, a little more than 1 billion years ago, what may have been the most important single event in the history of life took place: the formation of the first eukaryotic cells. (Some possible steps in this event have been alluded to in Chapters 6, 9, and 10.) Once the first eukaryotes were there, the evolutionary process started picking up at an ever-increasing pace. The first multicellular invertebrates arose some 800 million years ago, the first vertebrates about 200 million years later, and then, in accelerating succession, the higher fishes, the amphibians, the reptiles, and finally the mammals, which appeared approximately 300 million years ago. Primates are probably no more than 60 million years old, and their human offshoot branched out only 2 or 3 million years ago.

These are still very long time spans. Nevertheless, the speed at which evolution started moving once it discovered the right track, so to speak, and the apparently autocatalytic manner in which it accelerated are truly astonish-

This graph shows the rate of increase in the complexity of the genome as a function of evolutionary time. The shape of the curve suggests that complexity favored further complexity in some sort of autocatalytic fashion.

(Certain species diverge greatly from the figures shown because, for some ununderstood reason, their genome contains unusually large amounts of "ballast DNA.")

"God does not play dice" (Albert Einstein). "Our number came out in the game at Monte Carlo" (Jacques Monod).

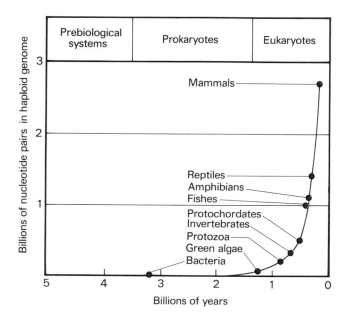

ing, especially in regard to the increasing generation time of the organisms involved, which should have slowed down the process. It took no more than 150,000 generations for an ape to develop into the inventor of calculus, whereas some 30,000 billion bacterial generations may have been needed to produce an amoeba. Which brings us back to the question of chance and design: How much chance? How much design?

The answer of modern molecular biology to this much-debated question is categorical: chance, and chance alone, did it all, from the primeval soup to man, with only natural selection to sift its effects. This affirmation now rests on overwhelming factual evidence. But it is not, as some would have it, the whole answer, for chance did not operate in a vacuum. It operated in a universe governed by orderly laws and made of matter endowed with specific properties. These laws and properties are the constraints that shape the evolutionary roulette and restrict the numbers that it can turn up. Among these numbers are life and all its wonders, including the conscious mind.

Faced with the enormous sum of lucky draws behind the success of the evolutionary game, one may legitimately wonder to what extent this success is actually written into the fabric of the universe. To Einstein, who once said: "God does not play dice," one could then answer: "Yes, He does, because He is sure to win." In other words, there may be a design. But it started with the "big bang."

Such a view is shared by some, not by others. The French scientist Jacques Monod, one of the founders of molecular biology and the author of the celebrated book *Chance and Necessity*, published in 1970, argued eloquently in favor of the opposite view. "Our number," he wrote, "came out in the game at Monte Carlo." And further: "The Universe was not pregnant with life, nor the biosphere with man." His final conclusion reflects the stoically (and romantically) despairing existentialism that greatly affected his generation of intellectuals in France: "Man now knows that he is alone in the indifferent vastness of the Universe from which he has emerged by chance."

This is nonsense, of course. Man knows nothing of the sort. Nor does he have any proof to the contrary, either. What he does know, however—or, at least, should know—is that, with the time and amount of matter available, anything resembling the simplest living cell, let alone a human being, could not possibly have arisen by blind chance were the universe not pregnant with them. Making such a statement does not in any way mean espousing a rigidly deterministic or, alternatively, a vitalistic view of the origin and evolution of life. It leaves full scope for essentially stochastic processes to operate behind these events, as conceived by modern Darwinian theory. But it does, by emphasizing the significance of the built-in constraints within which stochastic processes operate, provide a rational basis for a more optimistic philosophy than that of Monod.

There are many who have no need for such a philosophy and many more who cheerfully dispense with any sort of rational basis for the beliefs—or disbeliefs—they hold. Those, perhaps few in number, who are at the same time of a religious mind and respectful of scientific objectivity may derive some satisfaction from the view offered here. Surely they should find it more gratifying and uplifting than that of the so-called creationists, whose creed is really an insult to the Creator they believe in, making him into some sort of frivolously facetious, if not malicious, deity who has filled his work with innumerable molecular red herrings apparently designed for the sole purpose of leading scientists up some fantastically elaborate cosmic garden path.

Genetic Engineering

Certain genes, as we have just seen, are naturally endowed with a considerable degree of mobility. They can move from one place to another within a genome; they can jump in and out of a genome with the help of small satellite resting places, such as plasmids, bacteriophages, or viruses; finally, they can travel from one genome to another as passengers of these particles. Such migrations, which occur on a large scale in nature,

illustrate the "selfish" behavior of genes already referred to. Indeed, if the mechanisms involved are considered in a Darwinian fashion, the *raison d'être* for their selection often seems to be little more than the perpetuation of the genes concerned. The cells involved may or may not benefit from them, depending on the extent to which the wandering genes need healthy cells for their own survival and multiplication.

Seen in this context, the whole world of viruses and related particles appears as an intricate fleet of gene carriers, which probably originated in the first place through the evolutionary emancipation of mobile genes from complex prokaryotic or eukaryotic genomes. Once these facts became appreciated and scientists began to understand the mechanisms involved, the idea of using the fleet for the transport of certain specifically chosen passenger genes naturally presented itself. Nature provided the tools, and so genetic engineering was born, perhaps to become one of the most powerful techniques ever developed by mankind.

Indeed, the prospects evoked by genetic engineering appeared so awesome that scientists themselves took fright. In July, 1974, in an unprecedented action, a group of eminent scientists in the field declared a voluntary moratorium on further experimentation until its possible effects could be assessed. Predictably, this commendable manifestation of prudence and social responsibility created a tremendous stir and sparked a public debate of highly charged emotional content. Fears of uncontrollable, worldwide epidemics and of massive cancer invasions were raised. Accusations of "playing God," "tampering with the laws of nature," "crossing the sacred species barrier," were leveled. Prometheus, Pandora, the Sorcerer's Apprentice, Frankenstein, and other mythical figures were conjured to re-express mankind's deeply rooted distrust of new knowledge and technological innovation. Dark allusions were made to *Herrenvolk* philosophy and the manipulation of human heredity by unscrupulous rulers. A town council was even called to vote the outright banning of all DNA recombinant research from one of the world's most famous campuses. In a more sober vein, but yet responsive to the powerful currents stirred up within and without the scientific community, health agencies all over the world took over the problem and drew up elaborate guidelines that almost imposed more safeguards on the handling of genetically altered microorganisms than on that of the deadliest bacteria and viruses.

Today the furor has died down, or rather it has been replaced by another, of the gold-rush type. Companies have been formed to exploit the new technology, patents have been applied for, scientists have turned into entrepreneurs, and universities have become partners in commercial ventures, not without lengthy, soul-searching debates on the freedom of scientific information and on the disinterestedness of academic research. What, then, is the cause of all this turmoil? And, first, what is genetic engineering, known technically as artificial DNA recombinant technology, and what are its main purposes?

In theory—and now also in practice—what you do is simple. You take a piece of DNA and introduce it into a foreign cell in such a way that it will be replicated and, in some applications, also transcribed and translated. If only replication is sought, the manipulation is called molecular cloning, and the host cell is then invariably a bacterial cell chosen to produce as many DNA copies as possible in the shortest possible time. Cloning has no practical applications (except as a first step toward subsequent expression) but is an extremely powerful tool in basic research in that it can provide in almost unlimited quantities any segment of genetic information one may wish to analyze. In combination with other techniques of molecular biology, it is destined to allow a complete mapping of the genome of a cell, as well as a decoding of all the instructional elements that govern gene transcription and translation.

When actual expression of the DNA is aimed at, the objective may still be purely scientific, such as the elucidation of exactly what counts in a promoter sequence or what makes a gene oncogenic. But, in addition, two important types of applications are to be considered. One is large-scale manufacturing of a gene product, such as human insulin, interferon, antigens to be used for making vaccines, or any other rare protein that may be of practical value. The other application is genetic transformation of the recipient cell. Endowing certain crop plants with

the ability to utilize atmospheric nitrogen is one such aim that is currently attracting great interest. Correcting genetic diseases is another, though not yet within the realm of possibility.

A number of techniques are available for introducing the DNA into its recipient cells. It can be micro-injected, helped across the plasma membrane by special treatments of the cells, or enclosed within small membranous sacs or artificial phospholipid vesicles (liposomes, pp. 43–44) that fuse with the plasma membrane. As a rule, however, the yield of such procedures is low and their outcome uncertain. When applicable, the method of choice is to attach the DNA to the DNA of a vector, usually a plasmid or some viral particle, that happens to be naturally endowed with the appropriate means for introducing its DNA content into selected cells. This technique has the additional advantage that the vectors used often carry genetic markers, such as resistance to certain antibiotics, that render the recognition and isolation of the transformed cells very easy.

The tools for attaching the DNA to its vector are all borrowed from nature. They include restriction enzymes for cutting the DNAs at well-defined sites; nonspecific terminal deoxynucleotidyl transferases for fitting the vector and its passenger with sticky ends (e.g., poly-dG on one and poly-dC on the other), unless the restriction enzymes themselves are relied on to do this; DNA polymerase to fill in gaps; and DNA ligase to do the final stitching. The passenger DNA may be a simple piece of native DNA, a more complex mixture of such pieces, or the contents of a complete genome fragmented by some restriction enzyme. In the last case, the particular recombinant that one is interested in has to be fished out by special techniques from the highly motley population of transformed cells. Such experiments have been given the suggestive name of "shotgun experiments."

Quite often, the DNA is modified in various ways, either for the purpose of investigating the significance of some structural detail or in order to ensure its proper replication or transcription in the host cell. If the DNA comes from a eukaryote and the recipient cell is a prokaryote, the appropriate signals have to be given in

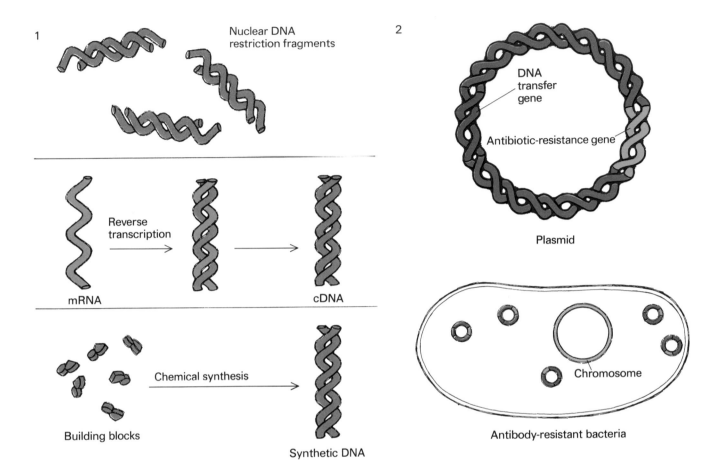

1

Nuclear DNA restriction fragments

mRNA — Reverse transcription → → cDNA

Building blocks — Chemical synthesis → Synthetic DNA

2

DNA transfer gene

Antibiotic-resistance gene

Plasmid

Chromosome

Antibody-resistant bacteria

prokaryotic language. The difficulty of crossing the prokaryote-eukaryote line is particularly great when expression of the transplanted gene is desired and splicing of the mRNA is required, as it is for most eukaryotic genes. Bacteria do not splice their own messengers; they lack the necessary machinery. To overcome this difficulty, the engineers use either reverse transcripts (cDNAs) of mature mRNAs or completely synthetic DNAs made to order according to the amino-acid sequence of the protein, natural or artificial, they wish the transformed cells to make. As can be seen, the possibilities of these new techniques are endless, and the amount of molecular juggling that can be accomplished by means of the currently DNA recombinant technique (genetic engineering).

1. The passenger DNA, to be cloned, may consist of fragments cut from the whole genome by means of restriction endonucleases ("shotgun" experiment), of cDNA transcribed from purified mRNA by means of reverse transcriptase, or of completely synthetic DNA.

2. The vehicle is usually a plasmid obtained from antibiotic-resistant bacteria and bearing one or more antibiotic-resistance genes, as well as genes that facilitate its transfer into its host. Sometimes a phage DNA is preferred and reinserted into the phage coat for transfer to the host (cosmid). With eukaryotic cells as hosts, rare plasmids or, occasionally, viruses may provide the vehicle DNA. Sometimes the DNA is introduced into the host cell without the help of a vehicle (transfection).

3

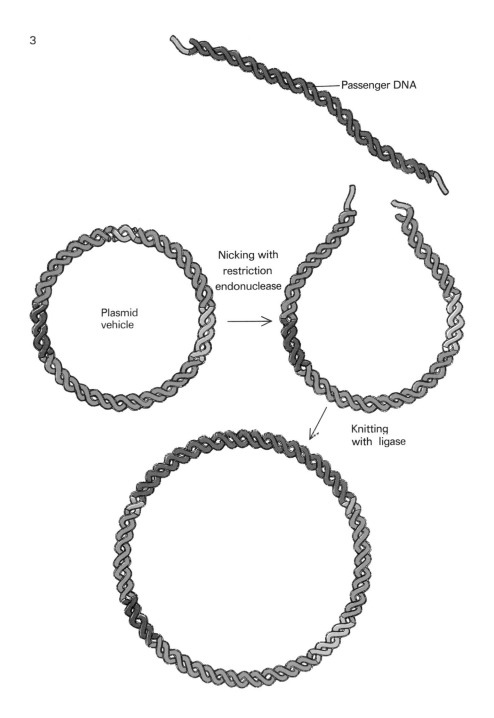

Passenger DNA

Plasmid
vehicle

Nicking with
restriction
endonuclease

Knitting
with ligase

3. To attach the passenger DNA to the plasmid vehicle, the plasmid must be nicked by a suitable restriction endonuclease and both passenger and vehicle must be fitted with complementary sequences (sticky or cohesive ends). Such will be the case if the passenger DNA has been fragmented with the same restriction endonuclease as has been used to open the plasmid (see the accompanying illustration, as well as that on p. 335). When passenger and vehicle are mixed, some molecules join by their sticky ends as shown, and the bonds can be sealed by a ligase. The drawback of this method is that many other combinations are formed in addition to the desired one. A more specific procedure is to fit the passenger DNA with one kind of ends and the vehicle DNA with the corresponding complementary ends. Sequences of dA for one, and of dT for the other, may serve the purpose. Alternatively, dC and dG may be used. This fitting is done with terminal deoxynucleotidyl transferase —an enzyme acting like DNA polymerase, but nonspecifically—and the appropriate deoxynucleoside triphosphate as substrate (dATP or dCTP for one and, correspondingly, dTTP or dGTP for the other). Then, when passenger and vehicle are mixed, only the desired combination can form, but there will be gaps at the junctions owing to the unequal lengths of the added ends. DNA polymerase is used to fill the gaps and the final knitting is then done by ligase action, as in the first case.

4

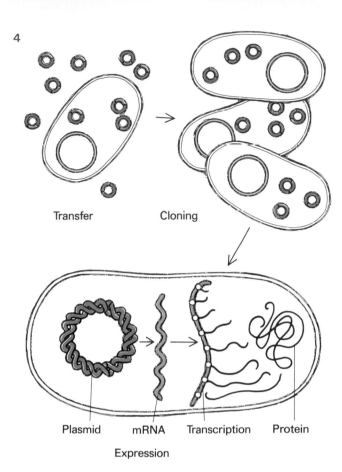

Transfer   Cloning

Plasmid   mRNA   Transcription   Protein

Expression

A genetically engineered giant. The mouse at the left is issued from an egg cell into which the human growth hormone gene, isolated by cloning and appropriately engineered for expression, was introduced by microinjection. The animal grew to twice the size of its control sibling.

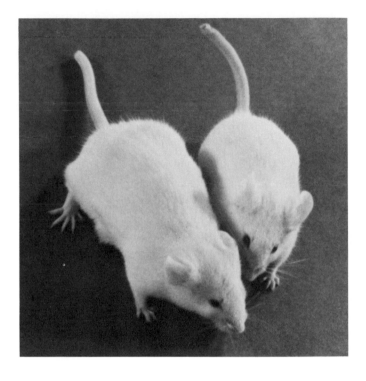

4. Exposure of the prospective bacterial (or eukaryotic) hosts to recombinant plasmids leads to the successful incorporation of the selected foreign gene into some of the host cells. If simple cloning of the gene (for the purpose of analysis, for instance) is desired, all that is needed is the correct replication of the plasmid and of its passenger in synchrony with cell multiplication. If the gene product is desired (industrial production of insulin or interferon, for example), engineering of the plasmid will have to include promoter and other regulatory sequences needed to induce the correct transcription and translation of the gene. A third application, not illustrated, involves the use of the modified cells themselves (cereals programmed to take up atmospheric nitrogen, correction of genetic defects, for instance). In all cases, the preliminary selection and cloning of the correctly changed host cells (usually a very small minority) is required. Advantage is taken of the antibiotic-resistance genes borne by the plasmid to select the cells that have taken up a plasmid. Various other "probes" are used to select the cells that contain the right gene in appropriate form. Such a selection is particularly stringent in experiments of the shotgun type, where thousands of different genes and gene fragments are handled simultaneously.

available recombinant kits is truly remarkable. There is no doubt that a new age has just started.

Bacteria are the obvious recipients for all cloning and manufacturing applications, in view of their rapid generation rate, ease of cultivation and selection, and large number of possible vectors. Some of the initial objections to genetic engineering arose largely from the fear that some bacteria, unwittingly transformed into highly pathogenic species, might escape into the environment. This risk appeared particularly hazardous because the most widely used bacterium is *E. coli*, a normal inhabitant of the human digestive tract. At first, strict physical containment of the facilities used was imposed as a safeguard against such conjectural accidents. Nowadays, reliance is put mostly on biological containment, itself a product of genetic manipulation (by conventional procedures). Mutant strains of *E. coli* have been produced that have strict temperature and nutritional requirements, absolutely incompatible with their survival in the human body or in any environment other than the highly specialized conditions provided in the culture vats.

The use of bacteria has, however, raised a number of difficulties, and eukaryotic hosts, especially the time-honored yeast cells, are also being developed. If transformation of a eukaryotic organism (for instance, a plant) is the final objective, then the host must, by definition, be of eukaryotic nature. Vectors for eukaryotes are less numerous and more difficult to manipulate than are bacterial vectors. Some of them have the additional disadvantage of being oncogenic, a fact that provided another reason for the original opposition to DNA recombinant techniques.

In certain important applications, vectors are dispensed with and the DNA is introduced into the cells by some other means. When the few cells that express the property that is sought are easily identified and cloned, this technique can be very powerful. It has played an important role in recent studies on cancerous transformation.

The Cancer Enigma

The Greeks saw it as a malevolent crab clawing deeply into its victim's flesh and gave it the crab's name of *karkinos* (*cancer* in Latin). They did not know how or whence it came. But they knew that once it has taken hold it never leaves off and just goes on eating away, spreading and growing to obscene proportions, until there is nothing left to consume. It attacks young and old indiscriminately but hits the old more frequently because of their longer exposure to its repeated assaults. And so it has become the main cause of death after cardiovascular disease, and certainly the most feared, in all areas of the world where removal of the primitive scourges of infection and starvation has left it enough scope to strike.

One reason why cancer is so frightening is its insidiousness. Not the slightest sign announces its onset, and it may remain silent and invisible for years while it establishes impregnable positions. All it needs is for a single cell to be transformed in such a way that it goes on dividing in undisciplined fashion (p. 320). This transformation appears to be irreversible and is in some way hereditary,

in that the cell's progeny share its refractoriness to population control. What causes the transformation? And what does it consist of? Those are the two main questions investigators are trying to answer, in line with the two great strategies of medical research: the one that treats the cell or patient as a black box and simply looks for causes in the hope that their knowledge may serve as an empirical guide to effective prevention; the other that tries to open the black box and to elucidate the disease-provoking mechanisms as a means toward rational cure or protection.

In 1760, a British surgeon, Sir Percival Pott, noted that chimney sweeps were peculiarly subject to cancer of the scrotum and deduced that soot retained between skin folds could be responsible for causing the disease. Since then, a number of occupational cancers have been recognized in workers exposed to such materials as radioactive minerals, tar, aniline dyes, asbestos, and vinyl chloride, and appropriate protective measures have been enacted. In addition, other potential causes of cancer have been looked for by widespread epidemiological investigations. This search has produced a wealth of statistical information linking the incidence of certain cancers to geographical factors, diet, smoking or drinking habits, and other forms of exposure either imposed by the environment or self-administered. In practice, however, except for some occupational cancers of restricted incidence, little progress in prevention has been achieved through this approach, either because the nature of the link is not yet sufficiently established to support a clear-cut recommendation or because the recommended measure turns out to be socially, economically, or politically unacceptable. A flagrant example of this is cigarette smoking, which is the most important single cause of cancer in the world, accounting for at least one-third of all human tumors, including some of the nastiest ones, such as lung cancer. Yet it is there for everyone to indulge in, while massive rallies are organized against purely potential hazards and interminable discussions center on the banning of this or that trace chemical accused of possibly causing one additional case of cancer in 100,000 inhabitants.

A valuable offshoot of the occupational and epidemiological approach has been the search for means of inducing cancer in experimental animals. It has allowed the identification of many carcinogenic agents, physical or chemical, and has produced important tools for the screening of chemicals, drugs, food additives, dusts, fumes, and other products or by-products of human industry. In addition, and perhaps more importantly, it has provided investigators with experimental models for the detailed study of the carcinogenic process itself.

An important notion revealed by these studies is that most carcinogens are mutagens, and vice versa. This discovery is clearly related to the hereditary character of the cancerous transformation mentioned above. Yet there is an important and puzzling difference between carcinogenesis and mutagenesis. Whereas a mutation occurs as an immediate consequence of the chemically induced genetic lesion, the development of cancer seems to be a multistep process. It generally requires repeated exposure to the carcinogen, and it is influenced by numerous factors, including the presence of substances called promoters or cocarcinogens that are not carcinogenic in themselves but enhance considerably the potency of carcinogens. No doubt genetic changes are involved in the transformation of a normal cell into a cancer cell. But the changes could be more complex or more subtle than ordinary mutations.

A revealing clue to the nature of these changes was discovered in 1911 by a young American scientist, Peyton Rous, who found that he could transmit a chicken sarcoma (a kind of cancer) to other chickens by injecting them with an extract of the tumor containing no living cells. The properties of the carcinogenic agent in the extract suggested that it might be a virus. As a rule, the scientific community behaves with the objectivity and open-mindedness that is expected from it. But occasionally it will greet with scepticism or disbelief, or simply ignore, a report that conflicts with some sort of consensus view. This happened to Peyton Rous's discovery. He did, however, live long enough to see his work vindicated and even recognized by a Nobel Prize—55 years after his finding was first reported. Had he lived to be a centenarian—he was 91 when he died in 1970—he would have been given an inkling of the truly momentous significance of his discovery, not only for our understanding of cancer, but for the whole development of science and technology.

Indeed, the Rous sarcoma virus has turned out to be the prototype of retroviruses, which, as explained earlier in this chapter, are those peculiar RNA-containing viruses that rely for multiplication on transcription of their RNA into cDNA by reverse transcriptase, followed by insertion of the cDNA into the host cell's genome (pp. 349–350). Thus the Rous virus served as a vehicle for the discovery of reverse transcriptase, one of the landmarks in the development of molecular biology. From the theoretical point of view, the existence of this enzyme contradicted the "central dogma" according to which genetic information is always transferred unidirectionally from DNA to RNA and from RNA to protein. (Note that the second part of the statement remains true and is likely to stay that way, as it amounts to a denial of the widely discarded Lamarckian view of the heredity of acquired characters). Reverse transcriptase has also provided genetic engineering with one of its most valuable tools and may, in this respect, be credited indirectly with many recent advances both in basic knowledge and in practical applications.

Another major product of Rous's discovery sprang from the genetic analysis of the virus. The oncogenic agent has a simple genome, made of only four genes. Of these, three carry all the information that is needed for multiplication of the virus; two code for the two viral proteins and one for the reverse transcriptase. The fourth gene, called the *src* gene, is in a way gratuitous. It has no function in multiplication of the virus but exerts a powerful effect on the host cell, being solely responsible for the cell's cancerous transformation. It is an oncogene—more correctly an *onc* gene—literally a cancer-causing gene. Some twenty such *onc* genes are known, each carried by certain kinds of retroviruses. They bear names such as *ras*, *myc*, *sis*, *erb*, *yes*, or *ski*—acronyms referring to the names of their viral carriers and of the diseases they cause.

The greatest surprise—and probably the most far-reaching gift of the Rous sarcoma virus to science—came after the DNA wizards arrived on the scene and started probing normal cells with cloned transcripts of the viral *onc* genes. They found that these genes do not, in fact, rightfully belong to the viruses that carry them, but rather were purloined from an infected cell in some earlier episode in the history of the viruses. Technically, as already mentioned (pp. 338 and 347), the phenomenon is known as transduction: a virus picks up a gene from its host cell and transfers it to another cell in the course of a subsequent invasion. Because of their peculiar mode of multiplication, which requires insertion of the viral genome into the host cell's genome, retroviruses are particularly well suited to act as transducing agents. Even so, the event remains a rare one: many retroviruses exist that do not carry an *onc* gene. To distinguish between the viral and cellular forms of the genes, the nomenclature uses the terms v-*onc* and c-*onc*, or, alternatively, oncogene and proto-oncogene, respectively.

As a cause of cancer, gene transduction is probably exceptional, but its discovery has brought to light an all-important clue: normal cells harbor genes that can cause a cancerous transformation under certain conditions (or after undergoing some modification). This started the sleuths looking for evidence implicating the same genes in the production of human cancers. And they have already uncovered plenty of it. At present, the whole field is in ferment. There are investigators hot on the trail in all parts of the world, and almost no week goes by without some new finding of significance being announced. The present account is written just before going to press. Yet, it will be outdated by the time it is published.

Two questions dominate the searches: What are the functions subserved by proto-oncogenes in normal cells? How are these functions subverted, or, otherwise put, how does a proto-oncogene become an oncogene? To answer the first question, the researchers have focused their attention on the proteins encoded by oncogenes and on their counterparts in normal cells. They have identified several. Initiates refer to them as p21, p53, pp60, which, if you know the cipher, often tells you almost as much about these proteins as their discoverers know of them: namely, that their molecular mass is 21, 53, or 60 kilodaltons (pp means that the protein is phosphorylated).

Some oncogene products, however, are beginning to be characterized. Among them is pp60, the product of the *src* gene. This protein has all the hallmarks of a key regulator. It is a protein kinase, belonging to a group of enzymes that we encountered before as playing a central role in the control of various enzyme activities (Chapters 13 and 14). Within this group, pp60 occupies a special position in that it phosphorylates tyrosine residues in its protein substrates, not serine or threonine residues, as other known protein kinases do. Its home is on the cytosolic face of the plasma membrane, where all the major transducing devices connected to surface receptors are situated (Chapter 13). Indeed, there are strong indications that pp60 belongs to a normal mitogenic system that is subject to stimulation by such growth factors as EGF (epidermal growth factor) or PDGF (platelet-derived growth factor). Remarkably, two other oncogene products have been traced to this system. One, the product of the *erb* gene, is related to the EGF receptor. The other, coded for by the *sis* gene, is a close relative of PDGF. It is of interest that these relationships were uncovered by computers programmed to deduce the amino-acid sequences of the proteins from the nucleotide sequences of the oncogenes that code for them and to match these sequences with those of proteins of known structure.

Exciting as they are, these findings do not, of course, signal the end of the trail. From the enzyme, we must now move to its substrate, and hence no doubt to new complexities in the chain of events underlying the form of mitogenic stimulation controlled by the tyrosine-phosphorylating pp60. Vinculin, which participates in the anchoring of actin microfilaments to adhesion plaques (Chapter 12), is a pp60 substrate and appears as a possible link in the chain. Adhesion plaques are dismantled in cancer cells and may have something to do with the phenomenon of contact inhibition mentioned at the beginning of Chapter 17.

Several other oncogene products, apparently unrelated to pp60, share with this protein the ability to catalyze the phosphorylation of tyrosine residues in certain proteins, suggesting the existence of a whole family of tyrosine-phosphorylating protein kinases that participate in

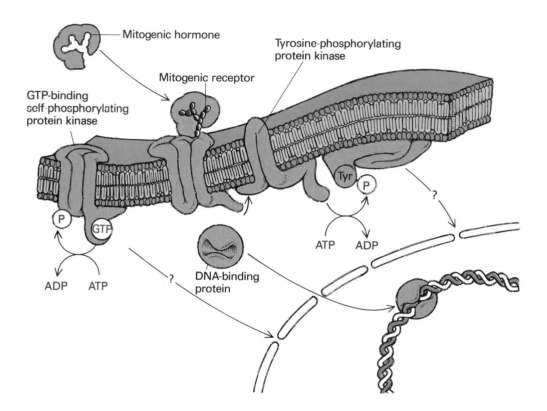

A schematic view of some proto-oncogene products and their presumptive function in normal mitogenesis.

growth control. A separate group of oncogene products, coded for by *ras* genes, comprises proteins that, like pp60 and its congeners, are bound to the plasma membrane and endowed with protein-kinase activity but that so far have been found to phosphorylate only themselves, and not on tyrosine residues at that; they have a strong affinity for GTP. A completely different family of oncogene products (e.g., of the *myc* gene) includes proteins that are situated in the nucleus, where they may bind to DNA.

The picture that emerges from all this is still very sketchy and hazy. But one important conclusion is already clear: many oncogenes—perhaps all—are derived from proto-oncogenes that code for a piece of some normal growth-controlling machinery. As to what causes a proto-oncogene to become a true oncogene, many differ-

ent mechanisms seem to be involved. One is mutation, resulting in an altered gene product. How frequently this happens is not known, but at least one case has been recognized in which a single point mutation accounts for the difference between oncogene and proto-oncogene. It concerns an oncogene of the *ras* family that has been extracted from a human bladder cancer (designated T24 or EJ). It differs from its normal counterpart by a single base-pair substitution, which results in the replacement of a glycine by a valine residue in position 12 of p21, the gene product. Replacement in the same position by an arginine or by a serine residue has been found in two murine *ras* oncogenes. In a number of other cases, the differences between oncogene and proto-oncogene are much greater, indicating multiple mutations, massive deletions, translocations, or other major changes. For example, the *erb* oncogene mentioned earlier seems to code for a maimed EGF receptor lacking, among other parts, the actual binding site for EGF, though not, however, the active mitogenic domain. This part of the molecule has presumably become unleashed as a result of the alteration and stimulates cell multiplication in uninterrupted, uncontrolled fashion.

Sometimes the cancer-causing change is quantitative rather than qualitative and consists simply in excessive expression of the gene. This may happen in many different ways: by gene amplification; by insertion of a strong promoter upstream of the gene, or of what is known as an enhancer of transcription in its neighborhood; by mutations that have the same effect; by various forms of transposition or recombination—including chromosome breakages and exchanges, which are prominent characteristics of several human cancers—resulting in translocation of the gene to a region where it becomes subject to such influences. The mechanism may even be posttranslational. The product of an oncogene carried by the DNA-containing polyoma virus is a protein that binds to pp60, presumably stabilizing it and thereby enhancing its expression. Incidentally, the oncogenes of DNA viruses are not stolen from host cells. They are authentic viral genes and play an essential role in virus multiplication (even though they do not code for viral proteins proper). Only in those rare cases in which the viral genome becomes inserted into the host-cell genome do these genes behave as oncogenes, as already seen.

In the gathering of all this new information, a technique known as transfection has rendered great services. Cells in culture are exposed to DNA under conditions such that some of the DNA enters the cells and, in a small number of them, becomes incorporated in the genome and undergoes normal replication and transcription. In this way, transforming DNAs can be recognized by the characteristic changes that they induce in the transformed cells and thus be isolated for cloning and sequencing. Interestingly, such experiments succeed only with established cell lines—the NIH 3T3 cells are a favorite—that is, with cells that have been "immortalized" by in vitro

culturing and no longer exhibit the phenomenon of programmed senescence typical of fully normal diploid cells. The resistance of normal cells to transformation by transfection can, however, be overcome if two distinct genes are used—for instance, one of the *myc* type, whose product goes to the nucleus and may be responsible for immortalization, and one of the *ras* type, which gives a plasma-membrane-bound product possibly responsible for the actual transformation. This finding provides an objective basis for the apparent multistep character of the cancerization process alluded to above.

In the intact animal, the successful establishment of a tumor and its ability to form metastatic colonies elsewhere in the body are subject to additional factors, some of which are genetic. In particular, mutations that block the expression of certain histocompatibility antigens on the surface of transformed cells (Chapter 3) may protect the cells against immune recognition and destruction by cytotoxic T cells and thereby precipitate the onset of a virulent neoplastic disease.

All in all, cancer researchers have reasons for satisfaction and optimism, especially considering that most of the new findings have been made in a span of less than five years. The cancer sphinx may not have met its Oedipus yet, but the day on which it does may not be too far away. What we will be able to do with the secret once we have it is another question. It is a tenet of the scientist's faith that understanding leads to control or, to quote Francis Bacon, that knowledge is power. But the example of genetic diseases, many of which we understand but do not yet control, shows that the road from understanding to control may not always be easy.

Interferon

Under the title "The IF Drug for Cancer," it made the cover of *Time* magazine before ever curing a patient. On the strength of very preliminary experimental evidence, millions of dollars have already been spent on clinical trials performed with the world's rarest, most expensive—$22 billion a pound until recently—and most unproved cancer drug. Future historians of science may well reflect on this strange phenomenon, which illustrates the increasing power that the public handling of scientific news is acquiring in our modern world over the directions that are given to scientific and medical investigation.

Let one thing be clear. Interferon is no quack remedy thrown on the market by some unscrupulous exploiter of human suffering. It is a natural substance of enormous interest, a prime agent in the defense of cells against viruses. The facts are as follows: When cells are attacked by a virus, the survivors, after a couple of days, become able to repel an attack not only by the same virus, but by many others. This phenomenon, called viral interference, is mediated by a protein that is specifically manufactured by the infected cells. That protein is interferon, of which there are several varieties, depending on the cells that make them.

As an antiviral agent, interferon may well have an important therapeutic future. The reason it is not in widespread use lies in its scarcity, itself the result of its specificity. To treat humans, you need human interferon. And to get human interferon—and in minute amounts, at that—you need human cells infected by a virus. But there is a bright side to the interferon coin. It happens to be one of the few proteins made from unspliced mRNA. This intriguing fact has allowed the successful transfer of the interferon gene in translatable form to bacteria by a shotgun type of experiment. If and when this laboratory finding is converted into a cheap industrial process, such irritating and often costly ills as the common cold, influenza, and herpes, as well as more serious conditions, such as hepatitis and viral pneumonia, could become things of the past, although the crucial tests still remain to be made.

In addition to its antiviral effect, interferon has a number of other biological properties, which include, at high enough dosage, the killing of certain tumor cells in culture and the stimulation of immune rejection mechanisms. These attributes, together with the possible involvement of viruses in the etiology of cancer, have sparked the first hopes that interferon may be an effective adjunct in the treatment of cancer. It is this cautiously raised possibility that somehow projected interferon into the limelight, to the point of draining considerable resources that many scientists felt might be used more profitably on less "iffy" and more cost-effective projects. Only time will tell whether this adventurous leap—an unusual event in the prudent and critical world of science—was justified.

19 | Making Two out of One: Mitosis and Meiosis

During our various digressions in pursuit of wandering genes, our nucleus has continued to copy its genome. Now all is quiet. The replication fever has died down. Even transcription is grinding to a stop. But it is an oppressive quiet, pent up, suffocating. The nucleus is literally bursting under its double load of chromatin and cracking at its seams. Something is bound to happen. Indeed, the cell is going over into the M phase. It is about to enter mitosis (Greek *mitos*, thread).

Prophase: a Major Packing Job and the Erection of a Giant Crane

The first intimation of things to come is the reappearance of the chromosomes. Remember how we saw them dissolve into a tangle of chromatin fibers when the nucleus first closed around us. Now we are witnessing just the opposite. The dispersed strings of nucleosomes re-spiral into regularly coiled ropes, which themselves roll and fold back to re-form the massive bodies that are such conspicuous components of the nucleus in dividing cells.

As we observe this remarkable sight, we get a better view of the anatomy of a chromosome than we had before. Its main component, you may recall, is the chromatin fiber, which is itself made by the helical coiling of a nucleosome-beaded string of DNA. Chromatin fibers are

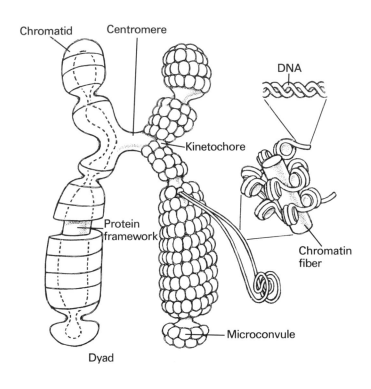

Model of metaphasic chromosome. Twin chromatids are linked by their centromeric regions to form a dyad. Each chromatid is supported by a framework of nonhistone proteins to which tightly bundled loops of chromatin (microconvules) are attached, possibly in a helical arrangement.

Chromatid

Centromere

DNA

Kinetochore

Protein framework

Chromatin fiber

Microconvule

Dyad

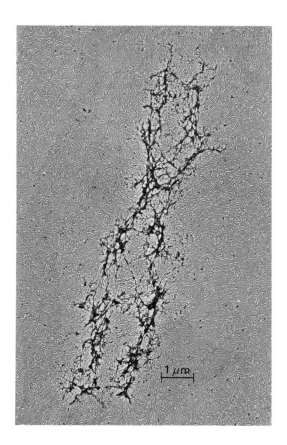

Electron micrograph of a metaphase chromatid dyad that has been depleted of histones. DNA loops attached to the protein framework are now completely unwound. Compare with the model at the left.

1 μm

about 25 nm thick, they have a pitch of about 10 nm, and contain some 400 nm of double-stranded DNA per helical turn. To form a chromosome, each chromatin fiber folds into a series of loops of 400- to 800-nm total length, which further bundle up into small, oblong balls, sometimes called microconvules. The resulting string of microconvules is itself garlanded around a network of nonhistone proteins that provides a structural skeleton to the chromosome. As we have seen, it is likely that the main lineaments of this arrangement persist around the matrix in the interphase nucleus, where the structure simply unfolds as a sort of multidimensional accordion to allow selected parts of the DNA to be transcribed. Lampbrush chromosomes give us some idea of how this might occur. What we see now is simply a refolding of the loops into microconvules, and their regrouping around the protein framework. We would dearly like to know how such an extraordinary packing operation takes place. Unfortunately, nothing can be distinguished through this impenetrable tangle. We can only admire the final result: thousands of genes and intercalating sequences neatly folded and coiled to about one eight-thousandth of their length into 46 (for a human cell) separate bundles, or, rather, pairs of bundles.

The DNA, remember, has just been replicated and now exists as twin copies, identical in virtually every single one of their six billion base pairs. In synchrony with this operation, freshly made histone molecules have entered the nucleus and supplied the replicating DNA with supplementary nucleosome cushions. From what now goes on before our eyes, we may deduce that cohorts of nonhistone structural proteins also came in and formed a second set of chromosomal frames for the new DNA strands to hang their nucleosome garlands on. Conse-

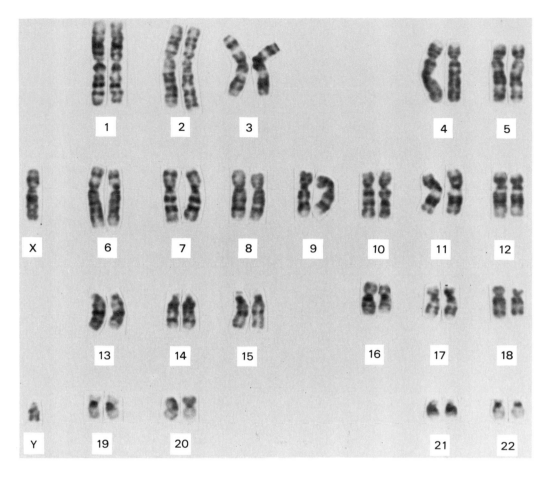

Karyotype of a normal human male. Metaphase chromosomes (note typical dyad structure) in a squashed cell preparation have been stained so as to reveal their banding patterns. The forty-six chromosomes in the photograph that was taken have been regrouped to construct this map.

quently, for each chromosome that we saw dispersing at the beginning of this visit, there are now two, in every way identical with each other.

The name chromatid is given to each member of such a pair. As they are seen to materialize during the first phase, or prophase, of mitotic division, the chromatids of each pair face each other like Siamese twins and are similarly interconnected by a bridge. Called the centromere, this junction exists between characteristically constricted parts of the chromatid bodies. It is a solid link, capable of resisting fairly harsh treatments, including the kind that need to be applied in order to squash cells and break open their nuclei. When dividing cells are squashed in this way on a glass slide and appropriately stained, the joined chromatid pairs, or dyads, appear under the microscope as typical X-shaped structures (sometimes Y- or V-shaped if the junction is eccentric), each of which can be identified by the length of its twin chromatids and by the position of the connection between them.

This technique, known as karyotyping, has become a valuable tool in genetic research and in the diagnosis of chromosome abnormalities. Thanks to special staining procedures, it has revealed an elaborate, longitudinal organization of the chromosomes into segments, or bands, which are distinguished by their degree of packing, sensitivity to thermic denaturation or proteolytic attack, ability to take up certain stains, or other physicochemical properties. Different bands are replicated at different times during the S phase and, as illustrated most vividly in the polytene chromosomes of diptera, correspond to different transcription domains (p. 317). More than a thousand bands have now been identified on the forty-six human chromosomes, and some five hundred specific genes have already been localized in them.

Chromosomes, we now clearly realize, are constructed in such a way that they can, without loss of their essential integrity, take up a protean variety of configurations. They can unwind and stretch out, severally or together, any of the thousands of individual segments of which they are composed, to the extent of adopting a completely dispersed state. Or, conversely, they can retract any of these segments, to the point of condensing fully into the compact packages that are needed to allow the genetic material to be moved around conveniently during mitosis.

Watching the enthralling spectacle of the mitotic metamorphosis of the chromosomes has been a rare privilege. But all the hustle and bustle it has entailed has hardly made our position more comfortable. Now, however, the pressure is beginning to ease, and feeble shafts of light come filtering in from the cytoplasm, announcing a new, remarkable event in the unfolding of the mitotic drama: the nuclear envelope is breaking open. With surprising rapidity, its continuous double sheet fragments into separate vesicles; the lamina falls apart; cytoplasm and nucleoplasm mingle their contents. But let us not rejoice too soon. We are not out of the nuclear woods yet. Replacing the padded wall of our prison is a huge, spindle-shaped cage made of hundreds of curved bars. Individual molecules and even small particles such as ribosomes readily pass between the bars, but not larger objects. Mitochondria remain outside, whereas chromosomes stay inside, as does our party. At least, we now have breathing space, and our view of the cell is almost unobstructed.

The cage that surrounds us was nowhere to be seen when we roamed through the cytoplasm, and it must have formed in direct connection with cell division. Indeed, it is none other than the mitotic spindle, so lovingly described by the first pioneers of cell exploration. They correctly saw it as playing an essential role in the forcible separation of the two chromosome sets during mitosis, but they could have had no inkling of how this extraordinary machine operates. Even today, our understanding of its mechanisms is fragmentary.

Construction of the spindle starts with one of the most mysterious phenomena of cell life: duplication of the centrosome, a small body situated near one of the poles of the

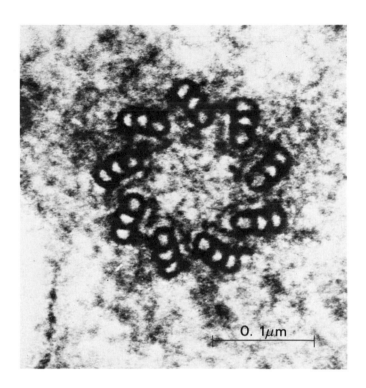

O. 1μm

nucleus. The centrosome consists of an amorphous matrix in which are embedded two short cylinders made of nine microtubule triplets and closely resembling the basal bodies, or kinetosomes, that form the roots of cilia (pp. 216–217). Called centrioles, these cylinders are characteristically oriented with their long axes at a right angle to each other. Such a pair of centrioles is called a diplosome. When the cell prepares to divide, the two centrioles move somewhat away from each other, and each grows a new centriole perpendicular to its flank to form two diplosomes, which organize further into centrosomes. How this occurs is not known. What follows is a little clearer, at least in certain cells. Nothing is more frustrating to the simple tourist in search of an "average" cell than the bewildering variety of mitotic events.

Starting from the two centrosomes, which apparently

act as microtubule organizing centers, microtubules start growing in all directions. Somehow, this growth process seems to push one of the centrosomes away from the other and to drive it around the nucleus down to the opposite pole. Eventually, long microtubules extend from one centrosome to the other, either continuously or in interdigitating fashion, all around the nuclear sphere and build the cage in which we found ourselves imprisoned when the nuclear envelope broke open. Shorter microtubules radiating from the centrosomes cap the two poles of the spindle with star-shaped crowns, the asters of classical cytology. And so the stage is set for the next phase, the metaphase.

Metaphase: Rigging the Chromosomes

After breakdown of the nuclear envelope, a second wave of microtubule assembly is set off inside the nuclear space, thanks to the influx of tubulin dimers from the cytoplasm. The organizing centers from which the microtubules grow are called kinetochores (Greek *kinein*, to move; khôra, place). They are situated on the main constriction that exists on each chromatid opposite the centromere. As a result of this disposition, each chromatid dyad sprouts, perpendicular to its main axis, two diametrically opposed sets of microtubules. At first, the direction of growth of the microtubules is variable because the chromatid pairs are more or less randomly oriented within the nuclear space. But as the microtubules grow longer, they align themselves parallel to the bars of the spindle cage, perhaps because of some constraint imposed by these bars.

The final outcome is the metaphase plate. The chromatid dyads are spread out (pushed by the growing microtubules?) in the equatorial plane of the spindle in such a way that this plane passes through their centromeres and the chromatids of each pair face opposing spindle poles (amphitelic disposition, see p. 381). The microtubules attached to them form two conical sheaves, each of which converges at one of the poles. The symmetry of this arrangement is remarkable and suggestive of some sort of balance of forces between the halves. The spindle itself, from being a simple, cagelike structure, is now filled with inner bars as well. These do not, like many of the cage bars, stretch all the way from one pole to another. They are always restricted to a single half-spindle and extend only from a kinetochore to the nearest spindle pole.

This elaborate rigging is completed by the addition of a number of poorly identifiable components, which may include short dynein arms connecting microtubules, as well as long actin fibers joined by myosin shafts. As we are about to see, the cell has just constructed a very special bidirectional chromosome hoist, all the more remarkable for being a purely temporary arrangement that will be dismantled immediately after having done its job, only to be reassembled at the next mitotic division.

Anaphase: Disjoining the Siamese Twins

We have come to the star turn of the mitotic show, one of the highlights of our tour: the anaphase, literally the ascension (Greek *ana*, upward). After what looks like a fair amount of pulling and wrenching, all the chromatid twins aligned on the metaphase plate become disjoined at about the same time and move away from each other as free chromosomes toward whatever spindle pole their kinetochore is attached to. It is a truly majestic sight, like watching 46 aerial gondolas lifting off in almost perfect synchrony over an unruffled lake in which their flight is mirrored in reverse. Their ascent is regally slow and may take from 5 to 10 minutes to cover the few microns (15–30 feet at our millionfold magnification) that separate them from a spindle pole. As the chromosomes rise, they are clearly lifted by their kinetosomes, while their hanging arms trail behind.

Many cell explorers have watched this display with awe and admiration, turning on it every instrument and technical device they could think of. Nevertheless, they are still debating about what does the pulling. That microtubules are involved is evident. In fact, the easiest way to "freeze" cells in midmitosis is to treat them with drugs, such as colchicine, vinblastine, or vincristine, that bind to tubulin dimers and inhibit their assembly. But where does the motive power come from? There are three possibilities, not necessarily exclusive of one another. One is "treadmilling" of microtubules—that is, their assembly at one end and disassembly at the other, such that any object attached sideways near the assembly end will be moved progressively toward the end where disassembly takes place. Another is "jacking" by means of dynein side arms attached to one microtubule and grappling onto an adjacent one, as in ciliary movement. Finally, there is the ubiquitous actin-myosin type of cytomuscle known to be involved in many forms of cell movement. The spindle contains parts of all three machineries, and all three may therefore be involved.

Another unresolved question is how the twin chromatids actually become separated. Are they wrenched away from each other mechanically? Or are their centromeres severed chemically by the action of some enzyme? Some of the movements we observe are suggestive of the mechanical explanation. But we may rightly wonder whether the tiny cytomuscles that do the pulling have the strength to break a connection capable, as we have seen, of resisting the brutal treatments that we inflict on the cells we prepare for karyotyping.

Amid all these uncertainties, some facts, at least, are beginning to emerge. It is clear that the kinetochore microtubules are directly engaged in the pulling mechanism, perhaps with the help of ATP-consuming cytomuscles of dynein type or actin-myosin type or both. As the microtubules pull their attached chromosomes, they shorten. Therefore, they are disassembled faster than they are assembled (if assembly takes place at all during this step, which is far from sure). This movement clears a widening space, essentially free of microtubules, between the two separating chromosome sets. For the first time since we entered a nucleus, we have space to relax. Except for the outer bars of the spindle cage, we could very well be in the cytoplasm.

In actual fact, we are not there yet. But the opportunity to get there is as good now as any we may encounter and we would be well advised to take advantage of it, even if it means bending or breaking a few microtubules. The spindle cage, you may have noticed, does not remain static while the chromosomes separate. It becomes both longer and narrower. The first change moves the poles further apart and thereby helps to widen the separation between the two sets of chromosomes. The second brings the microtubular bars of the spindle cage closer together and causes them to join progressively into bundles, called stembodies, in which they seem to be glued together by some amorphous material. As we shall see, these stembodies will be pressed even closer by the cleavage furrow. It is indeed time to leave. We may now watch the final phase, or telophase (Greek *telos*, end), from a safer vantage point in the cytoplasm.

Telophase: the Final Parting

By looking at either of the two sets of chromosomes after they have completed their ascent and become shorn of microtubules, we can follow from the outside the process of nuclear reconstruction that we saw earlier from the inside. Pieces of new lamina start forming around the clustered chromosomes while patches of endoplasmic reticulum converge upon them and fuse, leaving, however, the centrosome outside. Fenestrations form in this envelope and develop into pores. Inside, the chromosomes begin to disperse. Two typical interphase nuclei, each containing the same library of genetic texts, are the final outcome of these rearrangements, which obviously follow a complex script of instructions. Unfortunately, these are still totally undeciphered.

Occasionally, a cell may remain binucleated after mitosis. This occurs fairly often in the liver. As a rule, however, the cell undergoes division, or cytokinesis. In most animal cells, this happens by strangulation. A ring of plasma membrane, lined on its cytoplasmic face by hefty parallel bundles of actin-myosin fibers, is made to narrow

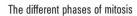

The different phases of mitosis

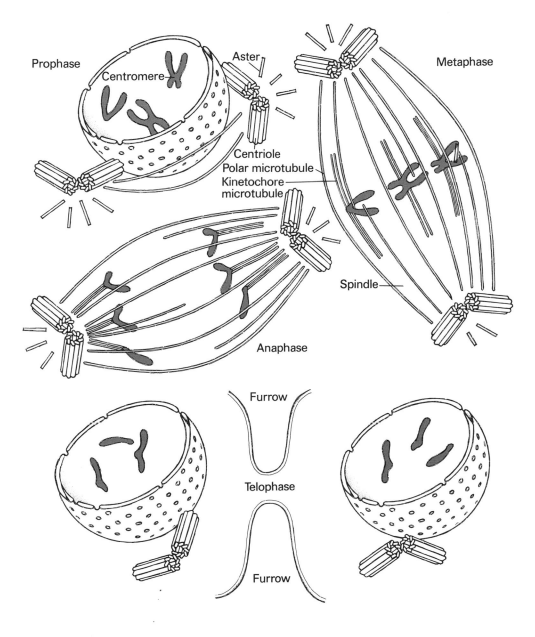

Prophase

Centromere

Aster

Metaphase

Centriole
Polar microtubule
Kinetochore
microtubule

Spindle

Anaphase

Furrow

Telophase

Furrow

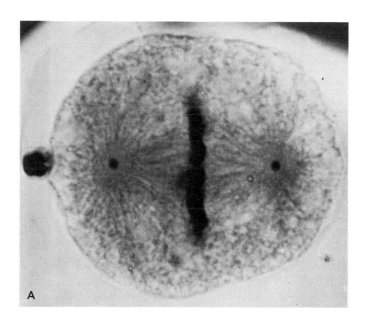

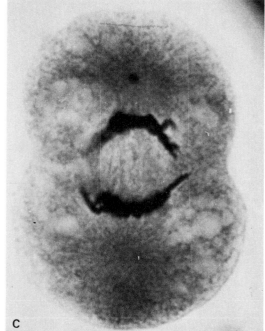

Mitosis in a fertilized egg of *Ascaris megalocephala* (a roundworm, parasite of the horse intestine). These preparations were made and stained in 1897 by J. B. Carnoy and photographed recently by L. Waterkeyn: (A) metaphase, transverse view; (B) metaphase, equatorial view; (C) anaphase.

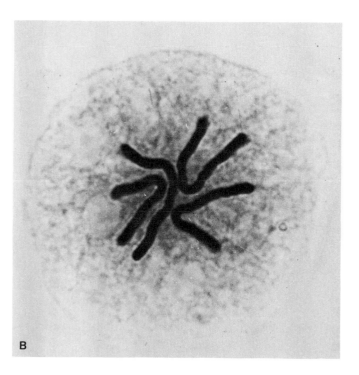

Cytokinesis.

A. Animal cell. A ring-shaped constriction (furrow) progressively narrows the cytoplasm between the two daughter nuclei, compressing the spindle microtubules into a tight bundle (stembody).

B. Plant cell. Secretion vesicles containing cell-wall material (phragmoplasts) align in the equatorial plane and fuse.

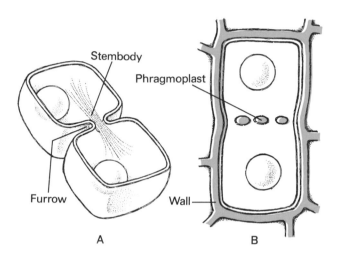

Stembody

Phragmoplast

Furrow

Wall

A

B

progressively into an ever-deepening furrow by the closing of this contractile noose. As the ring tightens, fluid cytosolic constituents are squeezed out on both sides, and the aggregated spindle microtubules (stembodies, p. 374) are pressed closer and closer together, eventually to form a single, dense shaft that finally becomes severed. When the lips of the constriction meet, they join by a typical *cis* type of fusion, and cell division is completed. The whole process has taken about an hour.

In plant cells, the mode of division is different. Only an incipient furrow is made by the plasma membrane, and the rest of the space between becomes filled with flattened vesicles originating from the Golgi apparatus. Called phragmoplasts (Greek *phragma*, partition), these vesicles fuse, also by *cis* fusion, to create a single flat cisterna that almost completely divides the cell into halves. Circular

fusion with the surrounding plasma membrane furrow, by what may be viewed as a concerted annular exocytic process, cuts this cisterna into two sheets, each of which becomes part of the plasma membrane of one of the daughter cells. The term "exocytosis" is quite appropriate here, for phragmoplasts are authentic secretion granules and their contents contribute to the formation of a wall between the two membranes. Often, small cytoplasmic bridges, called plasmodesmata, are spared by the fusion of the phragmoplasts and remain as direct connections between the two daughter cells.

Meiosis, or the Art of Getting Ready for Sex

A hen, it has been said, is only an egg's way of making another egg. Or, in the words of the nineteenth-century scientist August Weismann, *soma*, the body, serves to perpetuate the *germen*. But bodies do not arise out of single germ cells by parthenogenesis (Greek *parthenos*, virgin), but rather from two such cells by fertilization. Here is where sex comes in and, with it, meiosis (Greek *meioein*, to reduce).

When male and female germ cells, or gametes (Greek for spouses), unite upon fertilization, their nuclei combine into a single nucleus that contains the sum of the chromosomes contributed by each cell (p. 313). A diploid egg cell thus arises from the joining of two haploid germ cells. All the somatic cells that originate from this egg cell by successive mitotic divisions are likewise diploid. Should this be true also of the germ cells, the next generation would be tetraploid, the next one octoploid, and so on. This obviously would not do. There can be no sexual reproduction without some device for reducing the number of chromosomes in the germ-cell line. This device is meiosis.

To understand the mechanism of meiosis, we must remember that a diploid cell contains two complete haploid sets of chromosomes, one derived from the male, the other from the female germ cell. We will use the terms paternal and maternal to indicate the origin of any chromosome or gene, and we will call homologous the two chromosomes that bear the same genes or alleles in the paternal and maternal sets. Meiosis is a special kind of division that occurs uniquely in the maturation of both spermatozoa and ova. It results in the random segregation of one chromosome out of each homologous pair. As has been pointed out, even with intact chromosomes such a lottery would suffice to ensure a great deal of genetic diversity: 2^{23}, or more than eight million, different combinations for any human individual. But, in actual fact, diversification is immeasurably greater, thanks to crossing-over and the resulting recombination events that are characteristic of meiosis. Now that we are briefed, let us make a short excursion down the germ-cell maturation line to see how things actually happen.

In its early stages, meiosis is not very different from mitosis. DNA is replicated during interphase. And when prophase starts, the chromosomes recondense in the form of twin chromatids. From now on, however, things become very different. In meiosis, this prophase stage drags on for a very long time, sometimes many months or years—in the human female, oocytes go into meiotic prophase during the fifth month of fetal life and remain blocked at the diplotene stage (see p. 379) until the onset of puberty. The nuclear envelope remains intact during all that time, and the chromosomes uncoil into long, thin threads, which go through a great variety of contortions and changes in shape, whose main function is pairing, or synapsis, of homologous chromatid dyads into tetrads, followed by reciprocal exchanges of segments between paternal and maternal chromatids by recombination.

At the leptotene stage (Greek *leptos*, thin; *tainia*, tape), the chromatids are at their thinnest and longest, and they become attached by their extremities to specific points on the inner face of the nuclear envelope. Somehow, the matching ends of homologous chromatid pairs are brought close together by this process, so that the homologues become clearly distinguishable as coupled loops of various lengths. It is the zygotene stage. Then, starting from the attachment points, a synaptonemal complex builds up between the homologous chromatid pairs and "zippers" them intimately together into single thickened

loops in which the two components are no longer distinguishable. During this pachytene stage (Greek *pakhys*, thick), homologous stretches of paternal and maternal DNA bearing almost identical base sequences are brought into register with each other. The sole apparent function of this elaborate matching machinery is to allow the aligned partners to exchange homologous segments with each other (p. 337). The result is multiple recombination—of the "legitimate" type, of course, as befits the perfectly licit, and even obligatory, character of this kind of molecular intercourse. After a while, the tetrads are released from their attachment points on the nuclear envelope, the synaptonemal complexes disintegrate, and the homologous chromatid twins again separate, except where recombination has joined the DNA of a paternal chromatid to that of a maternal one. These junctions generate the characteristic chiasmata, or crosses, which revealed crossing-over to the early cytologists and opened the way to the interpretation of the phenomenon of genetic linkage. This final stage of meiotic prophase is called diplotene.

It is at this stage that meiosis often comes to rest, as already mentioned, especially in the female sex. Oocytes may, in the meantime, pursue their development and differentiation with the help of active transcription of their DNA. Their chromosomes then adopt the typical "lampbrush" shape, which results from the symmetrical unfolding and transcription of identical stretches on closely joined chromatid twins (Chapter 16). If you follow a lampbrush chromosome over a long enough distance, you are likely to come to a chiasma, from which you will find out that the dyad is really part of a tetrad.

Another phenomenon that takes place during the first meiotic prophase, actually at the pachytene stage, is the tremendous amplification of rRNA-coding genes in certain amphibian oocytes. Mention has been made of this selective replication process, which leads to the formation of hundreds of free nucleoli in a single nucleus and enables the cell to step up manifold its ribosome-manufacturing capacity (p. 302).

When this prolonged prophase finally ends, the subsequent events recall the similar stages of mitosis. A spindle is constructed around the nucleus in the cytoplasm; the

Prophase I

1. Leptotene

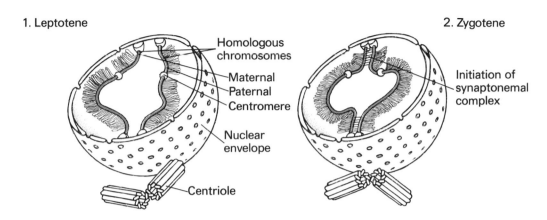

- Homologous chromosomes
- Maternal
- Paternal
- Centromere
- Nuclear envelope
- Centriole

2. Zygotene

- Initiation of synaptonemal complex

3. Pachytene

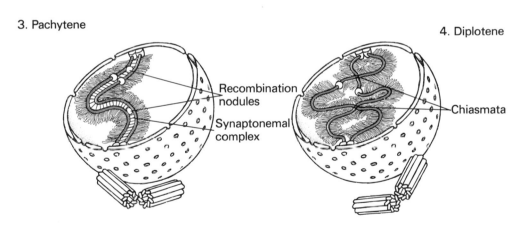

- Recombination nodules
- Synaptonemal complex

4. Diplotene

- Chiasmata

5. Diakinesis

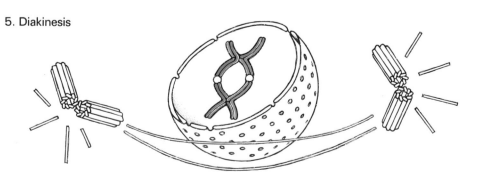

Details of prophase in the first division of meiosis.

1. Leptotene stage. Elongated dyads of closely apposed sister chromatids (not individually distinguishable at this stage) become attached to the nuclear envelope.

2. Zygotene stage. Homologous chromatid dyads of maternal and paternal origin are brought together in closely matched register by a proteinaceous synaptonemal complex linking the protein frameworks of the chromatids.

3. Pachytene stage. Synapsis is completed. Recombination (of the legitimate kind, Chapter 18) between homologous DNA loops takes place within recombination nodules.

4. Diplotene stage. Desynapsis has taken place. Dyads remain joined by one or more chiasmata at sites of recombination. Chromosomes have a typical lampbrush shape (Chapter 16).

5. Diakinesis. Chromosomes detach from the nuclear envelope and become progressively shorter and thicker. Their tetrad structure becomes visible.

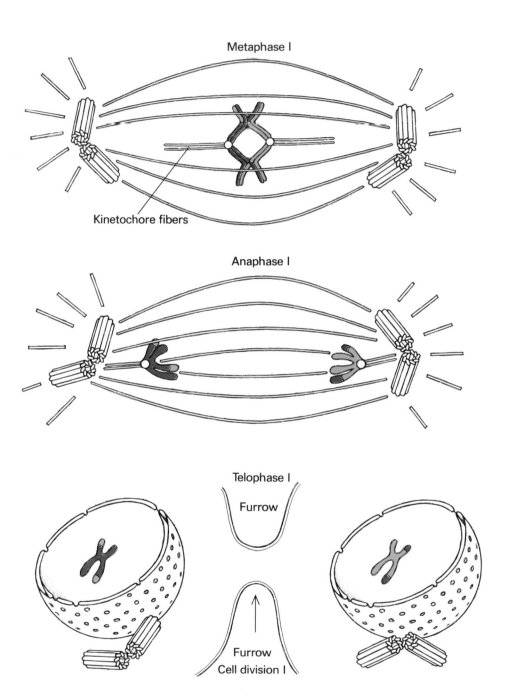

Metaphase I

Kinetochore fibers

Anaphase I

Telophase I

Furrow

Furrow
Cell division I

Details of metaphase, anaphase, and telophase in the first meiotic division.

Metaphase. The nuclear envelope disintegrates and a spindle forms. The tetrads line up in the equatorial plane in syntelic disposition (with the equatorial plane passing through the chiasmata).

Anaphase. Homologous dyads are pulled away from each other, with breakage of the chiasmata. Segments that have been exchanged by recombination accompany their new chromosomal owners.

Telophase. Two nuclei form and the cell divides. At this stage, each daughter nucleus contains a single set of dyads, some of which are of paternal, the other of maternal, origin (except for the segments exchanged by recombination).

nuclear envelope disintegrates; and microtubules assemble on the kinetochores. But there is one fundamental difference. Not forty-six dyads, but twenty-three tetrads come to be spread out on the metaphase plate. Furthermore, in their disposition, the equatorial plane does not cut through the centromeres and separate identical dyad twins (amphitelic disposition). Rather, the equatorial plane passes through chiasmata and separates paternal from maternal dyads (syntelic disposition). There are no rules for this separation, so that some paternal and some maternal dyads end up on each side of the plane. Furthermore, in many dyads, the twin chromatids are no longer in their original state or identical with each other, owing to the exchanges of homologous segments caused by crossing-over.

Because of the syntelic disposition, the subsequent anaphase severs chiasmata, not centromeres, and it separates two nonidentical sets of dyads, themselves made up of nonidentical twin chromatids. Genetic diversification has been achieved, but not yet the reduction in chromosome number. The two cells that form through this first meiotic division are still diploid. They will, however, go through a second division without an intervening S phase of DNA replication. In this second division, the chromosomes adopt the amphitelic disposition at metaphase, and the subsequent anaphase severs centromeres and separates twin chromatids, as in mitosis. But the number of dyads involved is only half that obtaining at mitosis, and their separation produces two haploid nuclei.

The mechanisms, the control, and the evolutionary origin of this extraordinary process are all beyond our present understanding. Its significance is of paramount importance. Meiosis is indissociably linked to sexual reproduction, which is itself considered the single most powerful generator of genetic diversity and therefore the most important agent of evolutionary experimentation and innovation. It is likely that, without sexual reproduction, evolution would have produced only very simple organisms. Even bacteria are known to indulge in some form of sexual activity. They differentiate into two distinct forms, which then conjugate and exchange genetic material, particularly plasmids.

Details of the second division of meiosis.

The two daughter cells of the first division divide again by a mechanism similar to normal mitotic division (amphitelic disposition and splitting of the centromeric bridge between sister chromatids at anaphase), except that division is not preceded by duplication of the DNA. Four haploid cells, each with a different and virtually unique complement of maternal and paternal genes, are produced.

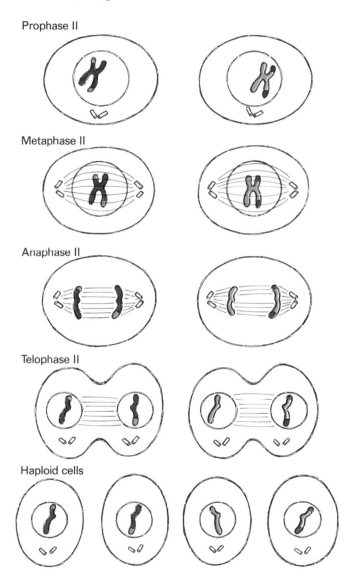

Prophase II

Metaphase II

Anaphase II

Telophase II

Haploid cells

Boy or Girl, Who Decides?

The answer to this question is: the father. (At least, if you are human; there are species in which it is the opposite.) But don't draw any sexist conclusions. The father has not the slightest control over his decision. It so happens that males, but not females, have one asymmetric pair of chromosomes. It is designated XY, as opposed to the XX combination that is characteristic of females. After meiotic reduction, one half of the mature spermatozoa carry an X chromosome, and the other half a Y chromosome. All ova, on the other hand, carry an X chromosome. It is easy to see that the sex of a conceptus will depend on whether an X-carrying or a Y-carrying spermatozoon fertilizes the ovum. Both are equally capable of doing so, and all the father can claim is the privilege of tossing a coin. Chance can, however, be influenced to the extent that X or Y spermatozoa can be separated and used selectively for artificial insemination. Animal breeders have attempted to do this, with some success.

Whatever vanity males may derive from their role in sex determination, it is more than offset by the price they pay for this distinction, which is the possession of only single copies of all the genes associated with the sex chromosomes. A woman carrying an abnormal gene on one of her X chromosomes may be rescued by a normal allele on the other X chromosome. But half her sons, on an average, will have only the bad gene. Hemophilia, a defect of blood coagulation, is the best known of these X-linked hereditary diseases that affect males and are transmitted by females. It has affected a number of members of the royal House of Hanover (Queen Victoria is the most famous carrier of the X-linked hemophilia defect).

Rather remarkably, females have chosen not to take full advantage of their superiority. At some stage in embryological development, their cells inactivate irreversibly one of their X chromosomes, which henceforth is seen as a highly condensed dot (Barr body). After this happens, both X chromosomes still undergo replication at each cell division, but only one is transcribed. Which one remains functional is determined by chance. The female organism thus consists of two types of cellular colonies, or clones, differing by the origin, paternal or maternal, of the X chromosome they express, and therefore by any X-linked character represented by different alleles on the two X chromosomes. There are many fascinating aspects to this genetic "mosaicism."

Making One out of Two

Cell fusion is the opposite of cell division. The most important such event is fertilization, the fusion of two gametes. It also occurs in a number of other processes, most characteristically in the development of muscle fibers. These arise from hundreds of embryonic muscle cells (myoblasts), which fuse to form a multinucleated syncytium. The fusion mechanism depends on a *trans* type of merger between the plasma membranes of the two partners. How it is triggered is not known.

However, ways have been found to make cells fuse in vitro. Such devices work not only with cells of the same type, as in the natural fusion of myoblasts, but also with cells of different types, even with cells from different species. This technique has become a powerful tool in the analysis of nucleocytoplasmic interactions and in the genetic mapping of chromosomes. It has also given rise to an extremely valuable procedure for manufacturing monospecific (monoclonal) antibodies.

Very early on our tour, we encountered the antibody-producing B lymphocytes, and we noted the important fact that each cell makes only a single kind of antibody. Any large-scale production of a given antibody is elicited by mitogenic stimulation of the few competent cells. These multiply into a clone of cells that all make the same antibody. There are numerous reasons why we would like to raise and maintain such clones in vitro and, thereby, to have available permanent antibody-producing factories of whatever specificity we need. We can isolate the B lymphocytes, but, unfortunately, we are unable to culture them. Those we get are postmitotic cells that do not divide.

The problem has now been solved by the artificial fusion of B lymphocytes with cancerous lymphoid cells

Monoclonal antibodies. Fusion of a nondividing antibody-producing cell with a dividing tumor cell gives rise to a hybridoma that combines the ability to produce antibodies with the capacity to divide. Descendants of a single hybridoma form a clone of cells that manufacture the same kind of antibody molecules (monoclonal antibodies). Such clones can be used for the industrial production of antibodies.

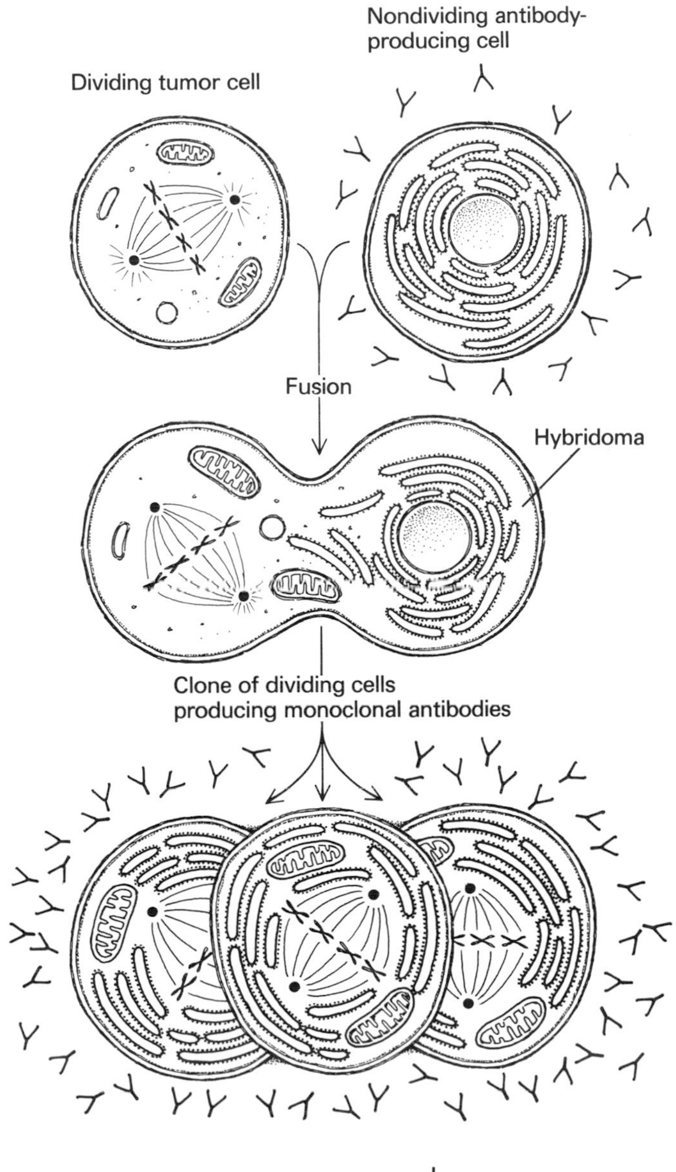

Dividing tumor cell

Nondividing antibody-producing cell

Fusion

Hybridoma

Clone of dividing cells producing monoclonal antibodies

(lymphoma cells), which possess the capacity of indefinite multiplication characteristic of the cancerous transformation. A small but significant fraction of the hybridomas obtained in this manner combine the antibody-producing capacity of the B lymphocytes with the multiplication potential of the lymphoma cells. By appropriate procedures, the desired clone can be isolated and used to set up a self-perpetuating antibody factory.

Final Exit

As promised, we have escaped from the nucleus, thanks to the disruption of the nuclear envelope during mitosis. It remains for us to find an exit from the cytoplasm. If we wish to respect the integrity of the cell, the only way we can get out of it is by budding.

Many viruses use this method. They are the membrane-wrapped viruses that provided us with our means of entry into the cytosol (p. 103). Such viruses actually acquire their membranous envelope when they bud from their original cellular birthplace. They do this in a very sophisticated fashion and carefully prepare their exit by instructing the cell to manufacture virus-specific membrane glycoproteins. Thanks to this specificity, the biogenetic pathway of the viral glycoproteins has been accurately traced. It begins in the rough-surfaced ER to which the nascent proteins are directed by a typical signal sequence and in which their glycosylation starts cotranslationally. After completion, the glycosylated proteins move on to the smooth-surfaced ER, probably by lateral diffusion, and are then transported to the Golgi apparatus, where they undergo further processing and glycosylation before being finally transported to the plasma membrane. In both transport steps, the viral glycoproteins move as parts of clathrin-coated vesicles. This information has provided much of the support for our belief that normal plasma membrane proteins follow the same biogenetic itinerary (Chapter 6).

While these events go on in the export department, viral nucleic acids and capsid proteins are made elsewhere in the cell, assemble into a complete nucleocapsid, and move near the cell surface, where they attract the virus-specific glycoproteins that have become inserted in the plasma membrane. Here, again, the virus has helped to throw light on important cellular mechanisms by showing how an object in the cytosol can induce the clustering of membrane-associated proteins. Presumably, these are mobile transmembrane proteins, capable of moving freely in the plane of the membrane by lateral diffusion and extending projections on both the outer surface of the plasma membrane—the carbohydrate "hairs," or "spikes," of the viral envelope—and its cytoplasmic face. Association between these inner projections and the nucleocapsid is what induces the clustering of the virus-specific glycoproteins.

Thanks to this mechanism, a patch of plasma membrane fitted with special viral components assembles in direct contact with the nucleocapsid and progressively surrounds it until the stem of the resulting bud snaps off by a *cis* type of membrane fusion. Thanks to this mechanism, which is illustrated on page 103, the virus is free to go and infect another cell using its membrane coat as means of ingress. Incidentally, some of these viruses, in particular a Japanese virus known as Sendai, can be made to try to invade two cells at the same time by merging simultaneously with the plasma membranes of the two cells. In doing this, they cause the two cells to fuse with each other. They provide one of the means used to create cellular hybrids.

It remains for us to copy the membrane-wrapped viruses to effect our final exit. By necessity, our technique will be cruder, since we cannot ask the cell to prepare tourist-specific membrane glycoproteins for us. In true human fashion, we will substitute ingenuity and expediency for the lack of built-in automatisms. Keeping in mind the fluidity, plasticity, and self-sealing properties of biomembranes, we will simply push gently but firmly against the plasma membrane. As expected, it yields and bulges under our pressure, flows and reshapes around us, until, as with the virus, a *cis* type of fusion lets us loose. This process seals the plasma membrane behind us and leaves the cell essentially intact, except for the hardly noticeable loss of a small patch of outer membrane.

This patch has ceased to be of use to its erstwhile owner. As for us, it could only help us to return into the cell. This, no doubt, we will want to do many times in the future, for much that could not be encompassed in this first visit remains to be seen, admired, and enjoyed in the living cell. Indeed, we will return. But, for now, the tour is over. You may take off your coats.

The Building Blocks of Living Cells

Little more than fifty different micromolecular building blocks account for the bulk of the organic matter found in any living organism. Diversity arises from the manner in which the building blocks are assembled into macromolecules, of which innumerable varieties exist, and, secondarily, from the activities of these macromolecules, in particular the enzymic proteins, which catalyze the formation of the thousands of small molecules that participate in metabolism and other specialized processes. The compendium that follows gives the structures of all major biological building blocks, together with those of most of the molecules mentioned explicitly in the book.

Carbohydrates

Monosaccharides

Glycoses. Carbohydrates are made of sugars, or monosaccharides. The simplest sugars, or glycoses, have the gross formula $(CH_2O)_n$. (Hence the name carbohydrate, which is actually a misnomer: carbohydrates are by no means hydrates of carbon.) Of the n oxygen atoms of a glycose molecule, $n - 1$ belong to alcohol groups, OH, and one to a carbonyl group, CO, which may be either aldehydic (in aldoses) or ketonic (in ketoses):

$$
\begin{array}{cc}
\text{H} & \text{CH}_2\text{OH} \\
| & | \\
\text{C=O} & \text{C=O} \\
| & | \\
(\text{CHOH})_{n-2} & (\text{CHOH})_{n-3} \\
| & | \\
\text{CH}_2\text{OH} & \text{CH}_2\text{OH} \\
\text{Aldose} & \text{Ketose}
\end{array}
$$

According to the number, n, of carbon atoms in their molecules, the glycoses are called trioses, tetroses, pentoses, hexoses, etcetera.

In glycoses, all the carbons bearing a secondary alcohol group (CHOH) are asymmetric—that is, such a carbon atom bears four different chemical groupings. These can be arranged around the carbon in two distinct, nonsuperimposable configurations, resulting in the existence of stereoisomers (Greek *stereos*, solid). Thus, glyceraldehyde (Chapter 7) can exist in two nonsuperimposable forms:

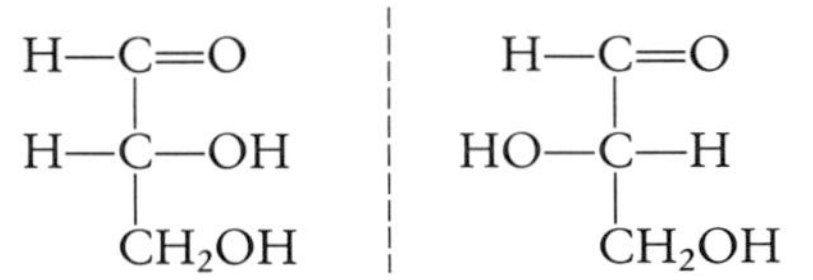

D-Glyceraldehyde   L-Glyceraldehyde

If there is more than one asymmetric carbon, the number of stereoisomers increases in exponential fashion. If the number of asymmetric carbons is represented by a, the number of stereoisomers is 2^a, forming $2^{(a-1)}$ pairs of enantiomorphs (Greek *enantios*, opposite; *morphê*, form), which are mirror images of each other. Each such pair is designated by a single name composed of a specific prefix followed by the suffix *ose*. The enantiomorphs are distinguished from each other by the letters D and L, depending on whether, in the planar projection of the molecular structure, the hydroxyl group on the penultimate carbon is situated at the right (as in D-glyceraldehyde) or at the left. For example:

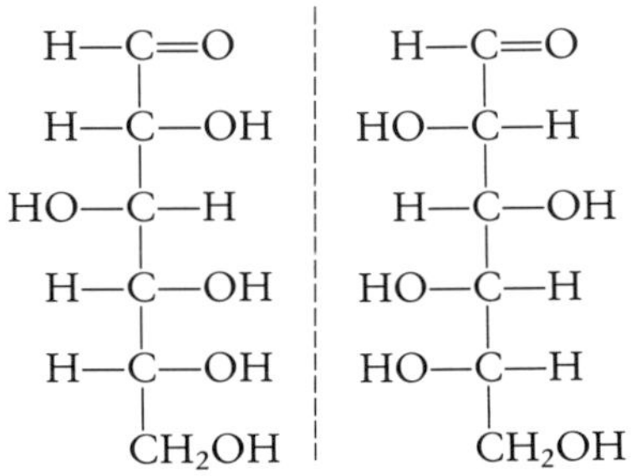

D-Glucose   L-Glucose

Sugars that have a chain of three or more carbons attached to the carbonylic carbon can undergo reversible cyclization by the addition of an alcohol group onto the double bond of the carbonyl group (hemiacetal bond). Two such cyclic structures can form without undue strain of the bond angles: the pentagonal furanose ring (4 carbons plus 1 oxygen) and the hexagonal pyranose ring (5 carbons plus 1 oxygen):

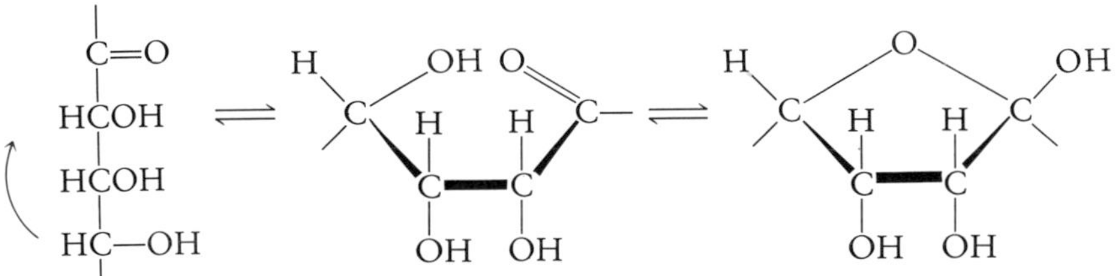

Furanose
(β form)

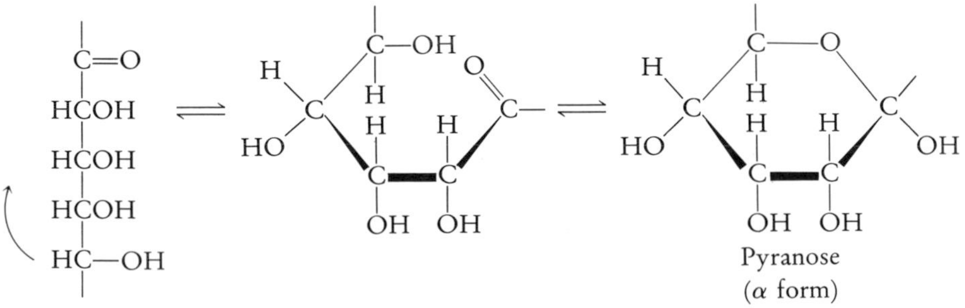

Pyranose
(α form)

As a result of this cyclization, the carbonylic carbon (carbon atom number 1 of aldoses, most often number 2 of ketoses) becomes itself asymmetric, so that two distinct cyclic stereoisomers exist of each linear variety. These are called anomers and are designated α and β, depending on the position of the anomeric hydroxyl with respect to the plane of the ring.

The cyclic forms of the sugars are of cardinal importance, as they are the real building blocks in the formation of all complex carbohydrates, thanks to their ability to combine by their anomeric carbon with a variety of molecules of carbohydrate or noncarbohydrate nature. Such combinations are called R-glycosides, or glycosyl-Rs. They are sometimes called hologlycosides if R is a carbohydrate, heteroglycosides if it is not. They can be of the α or β configuration:

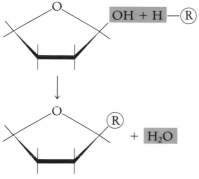

R-β-Furanoside, or β-Furanosyl-R

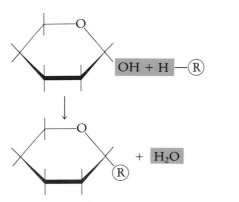

R-α-Pyranoside, or α-Pyranosyl-R

The structures of some of the most important glycoses follow, shown in their dominant cyclic configuration. Strictly speaking, deoxyribose, which lacks an oxygen atom in position 2, does not conform to the general formula of glycoses. Its structure is given here next to that of ribose because of the comparable functions of the two sugars in the formation of the two classes of nucleic acids.

PENTOSES

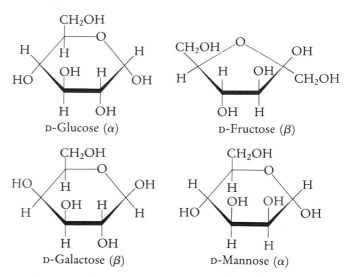

D-Ribose (β)   D-Deoxyribose (β)

(Color shows difference between the two pentoses.)

HEXOSES

D-Glucose (α)   D-Fructose (β)

D-Galactose (β)   D-Mannose (α)

(Color shows where configuration differs from that of glucose.)

Glycose derivatives. In addition to the glycoses, the monosaccharides include a number of related substances derived from the glycoses by removal of oxygen (deoxyglycoses); by oxidation, either of the aldehyde group in aldoses (aldonic acids) or of the terminal hydroxyl (glycuronic acids); by replacement of a hydroxyl by an amino group (glycosamines), which is generally acetylated (*N*-acetylglycosamines); by reduction (glycitols); or by some other modification. Except for the aldonic acids and glycitols, which have lost the capacity to cyclize by hemiacetal formation, these derivatives can form the same types of glycosides as do the glycoses from which they derive.

We have already encountered an example of a deoxyglycose in deoxyribose. Another is L-fucose, the 6-deoxy derivative of L-galactose, which is found at the end of some oligosaccharide side chains of glycoproteins. Among the aldonic acids, mention deserves to be made of ascorbic acid (vitamin C), which is a derivative of L-gulonic acid. The most important glycuronic acids are those derived from the three main aldohexoses—glucose, galactose, and mannose—by oxidation of the primary alcohol group in position 6. The most common glycosamines are derived from the same hexoses by substitution of a primary amino group for the hydroxyl in position 2. Representative examples are shown below.

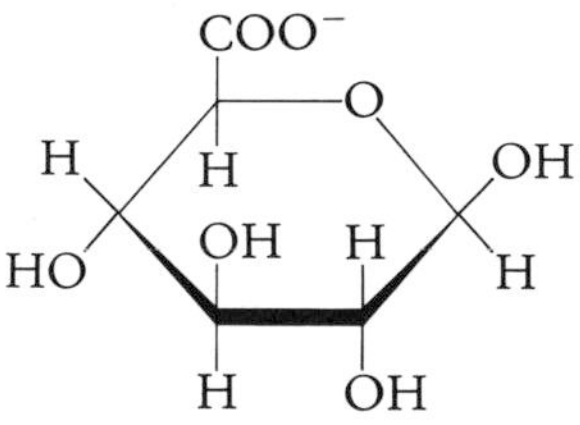

D-Glucuronic acid (β)   *N*-Acetyl-D-glucosamine (β)

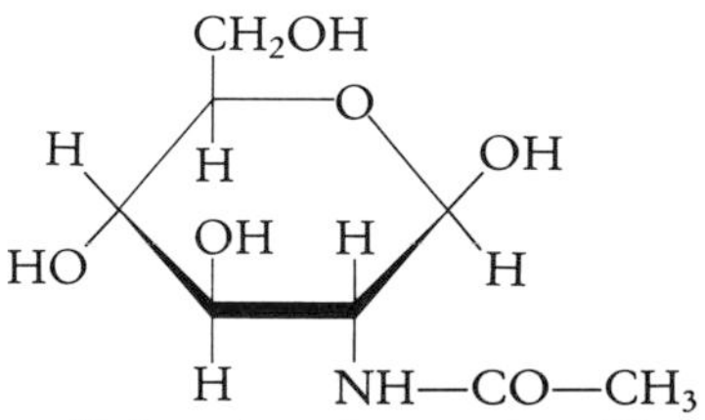

Among the more common glycitols are glycerol (reduced glyceraldehyde), a major component of lipids, and ribitol (reduced ribose), a constituent of flavin coenzymes (see p. 399). Related to the glycitols, except that they have only secondary alcohol groups, are the cyclitols, cyclic polyols obtained by isomerization of hexoses. *meso*-Inositol, found in certain phospholipids, forms in this manner from D-glucose:

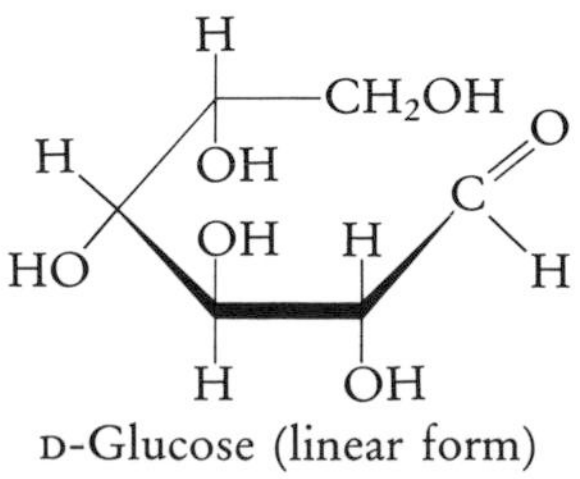

D-Glucose (linear form) 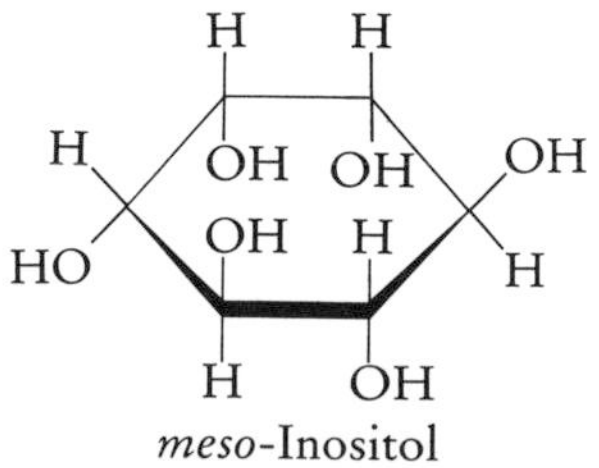   *meso*-Inositol

In this case, cyclization occurs by the addition of carbon-6 (instead of oxygen-5, as in pyranose formation, see above) to the carbonyl double bond.

Mention must finally be made of *N*-acetylneuraminic acid (NANA), or sialic acid, a molecule derived from the association of *N*-acetylmannosamine with pyruvic acid. Like fucose, it often occupies the end of oligosaccharide side chains.

Disaccharides

Sucrose, also named saccharose, is the ordinary sugar that we use for sweetening. It is made of α-D-glucopyranose and β-D-fructofuranose, linked by a rare form of mutual glycosylation.

Lactose, or milk sugar, consists of D-galactopyranose attached by a β-glycosidic bond to the oxygen in position 4 of D-glucopyranose.

Oligosaccharides

This group of substances includes a variety of linear and branched assemblies of up to fifteen sugar molecules, constructed mostly with glycoses and *N*-acetylglycosamines, often ending, as already mentioned, with either a fucosyl or a sialyl group. They form the side chains of glycoproteins and of glycolipids and thereby play a key role in the function of receptors (Chapters 3 and 13).

Polysaccharides

Nature, especially in the plant world, abounds in variously structured polymers of all the major glycoses. Starch, glycogen, and cellulose are all polyglucosides, or glucans. The first two have α-glucosidic linkages, whereas in cellulose the linkages are of the β-configuration. This apparently trivial difference accounts for the remarkable resistance of cellulose to degradation, commented upon in Chapter 2. Other homogeneous polysaccharides include mannans, galactans, and fructans, as well as polymers of pentoses, of glycuronic acids (pectins), and of *N*-acetylglycosamines (the chitin of crustacean shells).

A particularly important class of animal polysaccharides, sometimes referred to as mucopolysaccharides, is made of repeating disaccharide units usually consisting of an *N*-acetylglycosamine and either a glycose or a glycuronic acid. The hyaluronic acid of connective tissue (Chapter 2) is an example of such a molecule. It is made of *N*-acetyl-D-glucosamine and D-glucuronic acid. In many of these substances, as in most of the proteoglycans of connective tissue and in the anticoagulant heparin, some of the hydroxyls (occasionally, also, an amino group) bear a sulfuryl ester group. Such glycan sulfates are polyanions with a highly acidic character.

Heteroglycosides

The nucleosides (see p. 395) are the most important glycosides in which the R group is not a carbohydrate.

Proteins

Amino Acids

Proteins are made of L-α-amino acids, which, with one minor variant (proline, in which the R and NH$_2$ groups are linked), have the following structure:

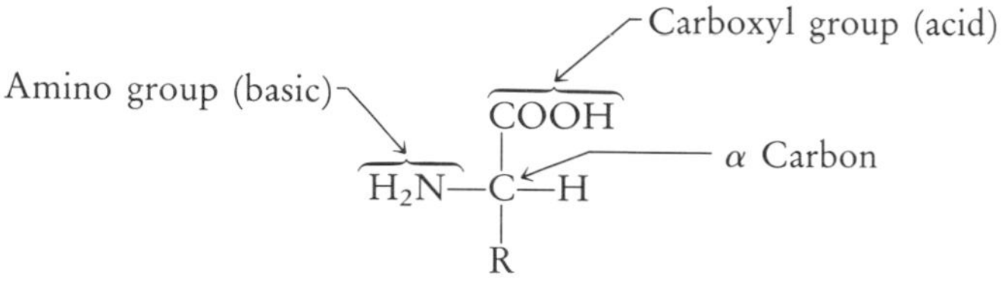

Amino group (basic) —↘   ↗— Carboxyl group (acid)

$$\overbrace{COOH}$$
$$H_2N—\underset{R}{\overset{|}{C}}{\overset{|}{\diagdown}}H \quad ——\ \alpha\ Carbon$$

Under physiological conditions (pH 7), amino groups are protonated for the most part, and carboxyl groups dissociated. Therefore, the most abundant natural structure of the amino acids is the amphoteric (Greek *ampho*, both), or zwitterionic (German *zwitter*, hermaphrodite), form:

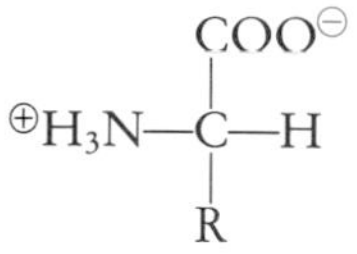

$$\overset{\ominus}{C}OO$$
$$^{\oplus}H_3N—\underset{R}{\overset{|}{\underset{|}{C}}}—H$$

Except in glycine, the α carbon is asymmetric, leading to the existence of two stereoisomers. They are designated D and L by analogy with the glyceraldehydes (p. 386):

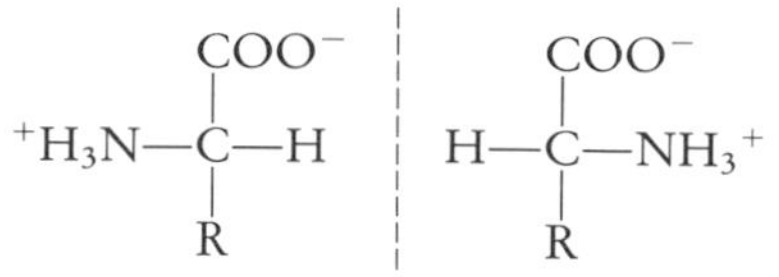

$$\begin{array}{cc} COO^- & COO^- \\ ^+H_3N—\underset{R}{\overset{|}{\underset{|}{C}}}—H & H—\underset{R}{\overset{|}{\underset{|}{C}}}—NH_3{}^+ \\ \text{L-Amino acid} & \text{D-Amino acid} \end{array}$$

Only L-amino acids are found in proteins, but their D stereoisomers are components of the bacterial cell wall and other bacterial products, including some antibiotics.

As explained in Chapters 2 and 15, twenty distinct amino acids are genetically coded and serve as building blocks for the synthesis of proteins. A few more—the hydroxyproline of collagen is an example—arise by posttranslational modification. The structural formulas of the genetically coded amino acids are given on the next page, arranged approximately in order of increasing hydrophilicity, a character that plays a dominant role in the folding and associative properties of proteins (Chapters 2 and 3). The critical importance of the physical properties of the R groups for the conformation and biological activity of proteins is illustrated by the structure of the genetic code. As pointed out in Chapters 15 and 18, the code seems such as to minimize the proportion of point mutations that lead to the replacement of amino-acid residues by residues of significantly different physical character. Most likely, the genetic code acquired this particular structure in the course of evolution by the operation of natural selection.

Peptides and Proteins

Amino acids join by amide bonds (called peptide bonds in this particular case) to form linear assemblies, or peptides. Except for a number of special oligopeptides, mostly hormones, the vast majority of natural peptides are macromolecules containing up to several hundred amino-acid residues. When fully extended, these polypeptide chains have a zigzag shape, with the R groups alternating on each side of the backbone:

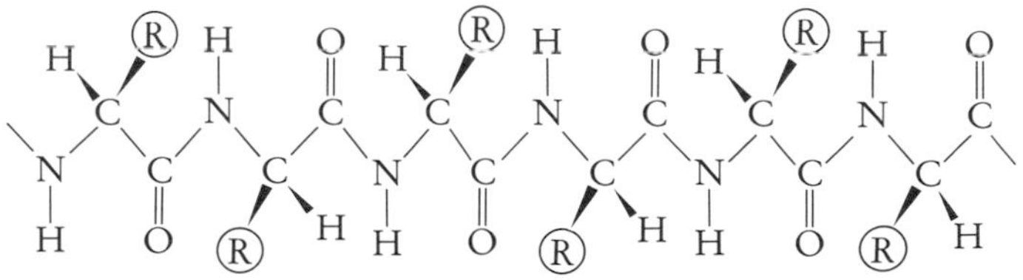

This basic structure may then undergo further coiling and folding, to create the convoluted three-dimensional forms adopted by most proteins. Examples of such structures have been considered in Chapters 2, 3, 12, and 15, among others.

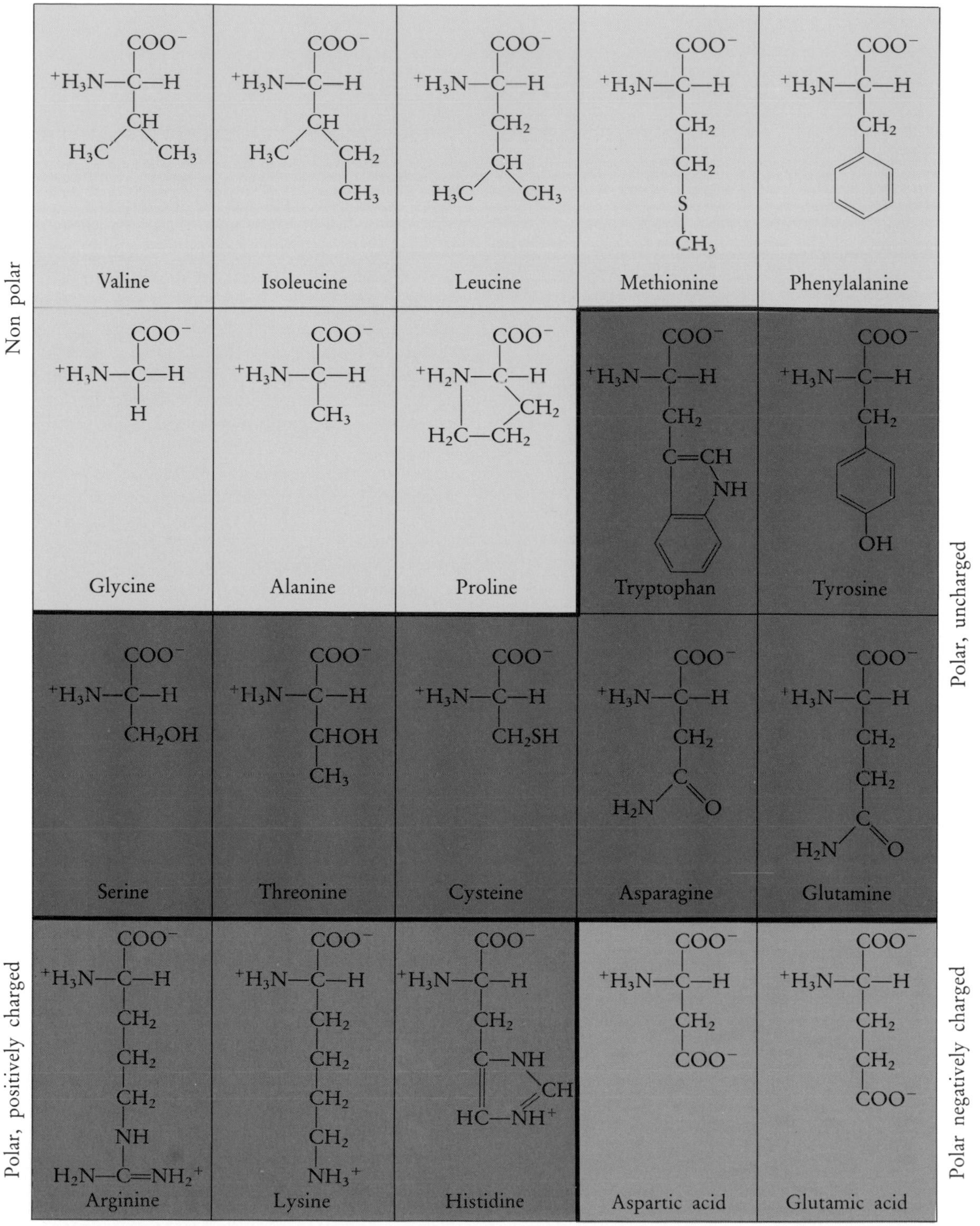

Non polar

Polar, uncharged

Polar, positively charged

Polar negatively charged

Valine	Isoleucine	Leucine	Methionine	Phenylalanine
Glycine	Alanine	Proline	Tryptophan	Tyrosine
Serine	Threonine	Cysteine	Asparagine	Glutamine
Arginine	Lysine	Histidine	Aspartic acid	Glutamic acid

Valine: $^+H_3N-C-H$, COO^-, CH, H_3C, CH_3

Isoleucine: $^+H_3N-C-H$, COO^-, CH, H_3C, CH_2, CH_3

Leucine: $^+H_3N-C-H$, COO^-, CH_2, CH, H_3C, CH_3

Methionine: $^+H_3N-C-H$, COO^-, CH_2, CH_2, S, CH_3

Phenylalanine: $^+H_3N-C-H$, COO^-, CH_2

Glycine: $^+H_3N-C-H$, COO^-, H

Alanine: $^+H_3N-C-H$, COO^-, CH_3

Proline: $^+H_2N-C-H$, COO^-, CH_2, H_2C-CH_2

Tryptophan: $^+H_3N-C-H$, COO^-, CH_2, $C=CH$, NH

Tyrosine: $^+H_3N-C-H$, COO^-, CH_2, OH

Serine: $^+H_3N-C-H$, COO^-, CH_2OH

Threonine: $^+H_3N-C-H$, COO^-, $CHOH$, CH_3

Cysteine: $^+H_3N-C-H$, COO^-, CH_2SH

Asparagine: $^+H_3N-C-H$, COO^-, CH_2, C, H_2N, O

Glutamine: $^+H_3N-C-H$, COO^-, CH_2, CH_2, C, H_2N, O

Arginine: $^+H_3N-C-H$, COO^-, CH_2, CH_2, CH_2, NH, $H_2N-C=NH_2^+$

Lysine: $^+H_3N-C-H$, COO^-, CH_2, CH_2, CH_2, CH_2, NH_3^+

Histidine: $^+H_3N-C-H$, COO^-, CH_2, $C-NH$, CH, $HC-NH^+$

Aspartic acid: $^+H_3N-C-H$, COO^-, CH_2, COO^-

Glutamic acid: $^+H_3N-C-H$, COO^-, CH_2, CH_2, COO^-

In the table on the facing page, the structures of the twenty genetically coded amino acids are arranged in the approximate order of increasing hydrophilicity. The first horizontal row, which includes valine, isoleucine, leucine, methionine, and phenylalanine, comprises the most hydrophobic amino acids; they tend to be buried in regions from which water is excluded. The next row contains amino acids that, in spite of essentially hydrophobic R groups, are less strictly segregated from water—glycine and alanine, thanks to the small size of their R groups; proline because of the rigidity imposed on its molecule by the cyclization of the R group with the primary amino group; and tryptophan and tyrosine, owing to their possession of a polar group (NH or OH) that can participate in hydrogen bonding. In the third row, serine and threonine (with their OH group), cysteine (with its SH group), asparagine and glutamine (with their CO and NH_2 groups) all have a clearly hydrophilic character. Their R groups are, however, electrically neutral and can be accommodated also in hydrophobic regions. Cysteine, in addition, can join to another cysteine by a disulfide bridge, S—S, which has an essentially nonpolar character. The last row contains the five amino acids that have an electrically charged R group. Strongly hydrophilic, they are confined to hydrated areas. Arginine, lysine, and histidine are basic and have a positively charged R group. (Only about half the histidines, on average, are so charged at any given moment.) Aspartic and glutamic acids are negatively charged.

Lipids

Fatty Acids

The main building blocks of lipids are long-chain fatty acids with an even number of carbon atoms, mostly sixteen and eighteen. Some are fully saturated, as is the C_{16} palmitic acid:

$$CH_3-(CH_2)_{14}-COOH$$

Others are unsaturated. In the C_{18} series, for example, oleic acid has one, linoleic acid two, and linolenic acid three double bonds. A specially important polyunsaturated fatty acid is arachidonic acid, with twenty carbon atoms and four double bonds. It gives rise by oxidation to the prostaglandins, a family of molecules of high biological activity.

The double bonds in natural fatty acids are mostly of the *cis* type, with the result that the carbon chain is bent at an angle

where they occur. Such kinks are of obvious importance for the organization of lipid bilayers (Chapter 3). The alteration in molecular shape caused by double bonds can be appreciable, as shown by the following comparison between oleic acid and its saturated C_{18} congener, stearic acid:

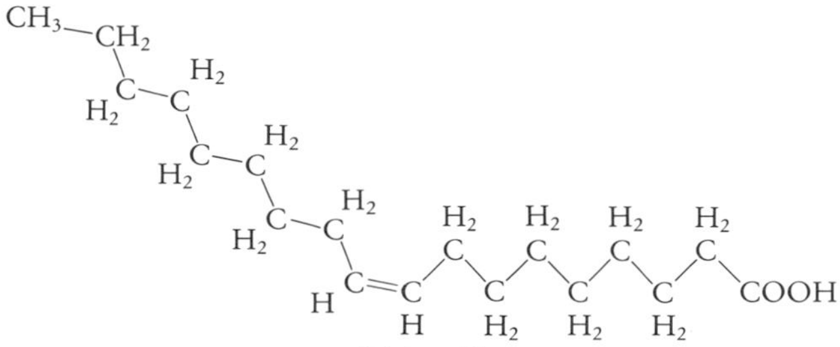

Oleic acid

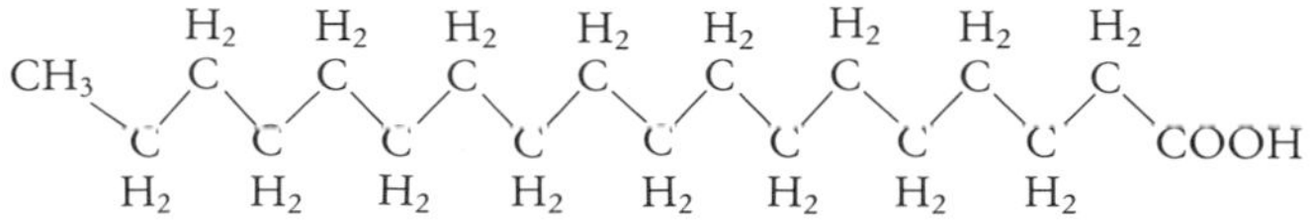

Stearic acid

As indicated in Chapter 3, the fatty acids provide many of the long hydrophobic tails characteristic of most lipids. In free form they are fully dissociated under physiological conditions and thereby acquire the hydrophilic, negatively charged head to which soaps owe their amphipathic character. Mostly, however, fatty acids occur as compounds of lipids, in which their carboxyl group is engaged in an ester or amide linkage.

Neutral Lipids

Triglycerides. Triglycerides are fatty acyl esters of the trialcohol glycerol, the glycitol derived from glyceraldehyde (see pp. 386 and 388):

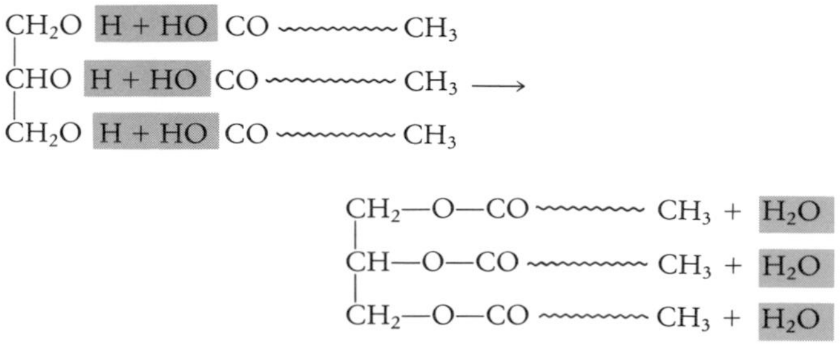

Such molecules have almost no hydrophilic character and are completely immiscible with water. They constitute the bulk of animal and vegetable fats and oils. They fulfill essentially no structural function, but have the advantage of providing the highest calorie count (energy of oxidation) per unit of weight of all biological constituents. Mostly, they are stored to serve as fuel.

Waxes. These are fatty acyl esters of long-chain alcohols, sometimes of very great length—the myricylic alcohol of beeswax contains as many as thirty carbon atoms. Except for the ester group, waxes are structurally very close to the high-molecular-weight hydrocarbons that make up petroleum jelly. Like such alkanes, they are extremely hydrophobic.

Phospholipids

Except for sphingomyelin (see p. 393), the phospholipids, which make up the lipid bilayers essential to the structure of all biomembranes (Chapter 3), are derivatives of phosphatidic acids. These may be viewed as triglycerides in which an external fatty acyl group has been replaced by a phosphoryl group:

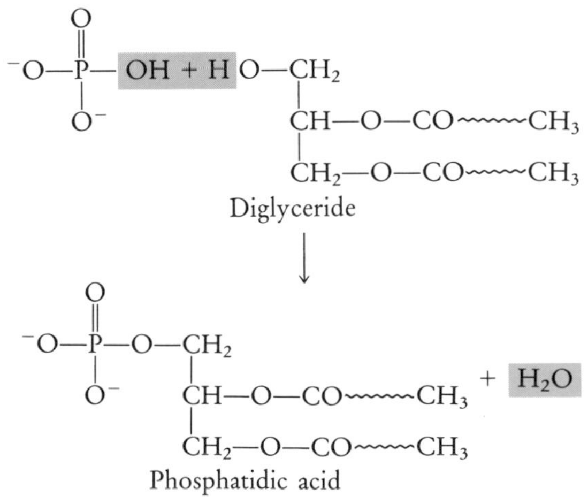

Diglyceride

Phosphatidic acid

Mostly, this phosphoryl group forms a second ester bond with some alcohol (represented by ROH):

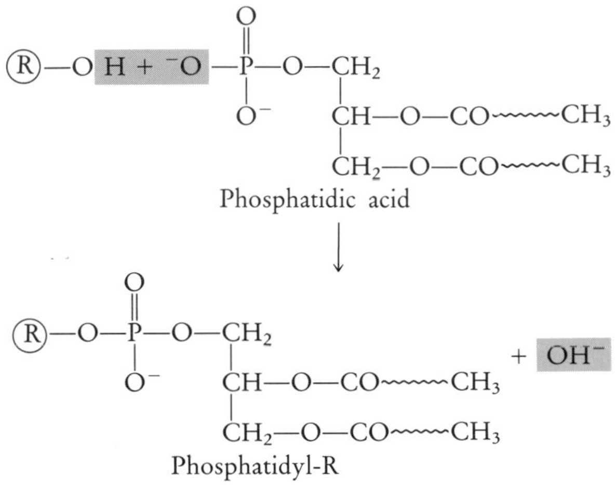

Phosphatidic acid

Phosphatidyl-R

Among the alcohols participating in such associations are glycerol itself and some of its combinations, the cyclic hexitol inositol (p. 388), and especially the aminoalcohol ethanolamine, its methylated derivative, choline (also a constituent of the neurotransmitter acetylcholine, Chapter 13), and its carboxylated derivative, the amino acid serine:

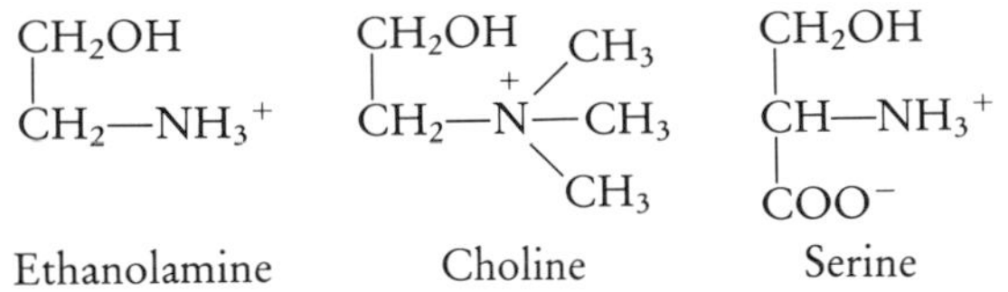

Ethanolamine   Choline   Serine

These are all highly polar molecules, which, together with the phosphoryl group, give the phospholipids their hydrophilic heads. Here, for example, is the structure of phosphatidylcholine, known earlier as lecithin (Greek *lekithos*, egg yolk), a major constituent of all biomembranes:

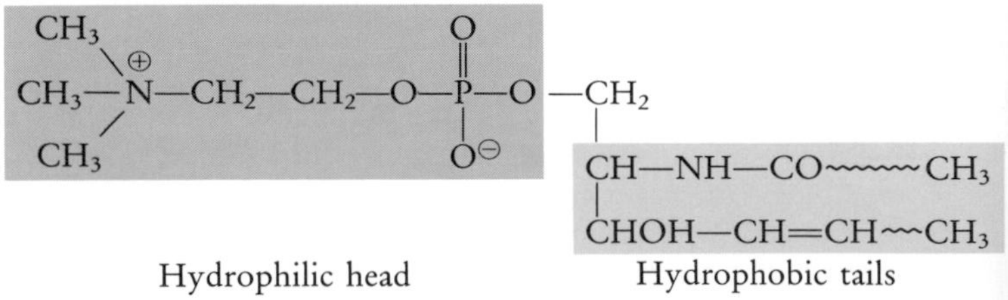

Hydrophilic head   Hydrophobic tails

Sphingolipids

This diverse group of complex lipids is composed of derivatives of sphingosine, a long-chain aminoalcohol with the following structure:

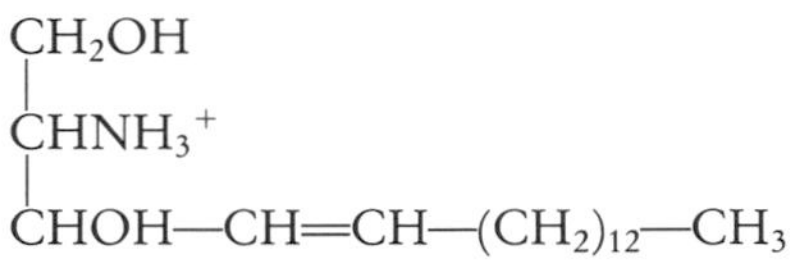

$$CH_2OH$$
$$|$$
$$CHNH_3^+$$
$$|$$
$$CHOH—CH=CH—(CH_2)_{12}—CH_3$$

As a rule, sphingosine has a long-chain fatty acid attached to its amino group by an amide linkage. Called ceramide, this combination resembles diglycerides in having a two-pronged hydrophobic tail and a free hydroxyl to which various groups can be linked. The diverse family of sphingolipids arises in this manner. In association with phosphorylcholine, for example, ceramide forms sphingomyelin, a phospholipid very similar in shape and in physical properties to phosphatidylcholine:

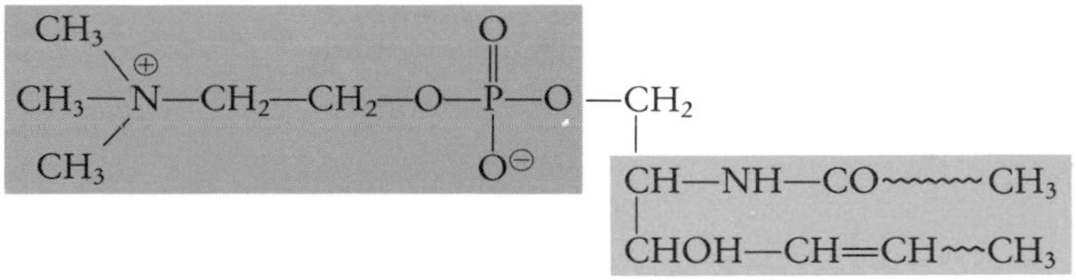

Replacement of the phosphorylcholine group in sphingomyelin by a β-D-galactosyl group gives a cerebroside:

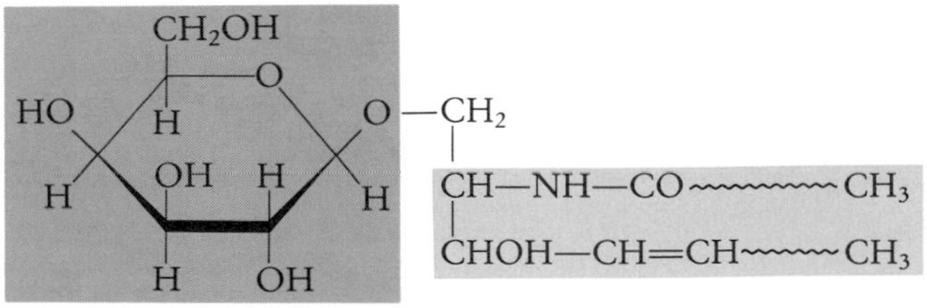

With glucose instead of galactose, a glucocerebroside is obtained. These substances, in turn, give rise to a variety of derivatives, including the sulfocerebrosides, in which some of the hydroxyl groups of the sugars are esterified by sulfuric acid, and the gangliosides and other complex glycolipids, in which oligosaccharide assemblies are attached to the ceramide, usually by a terminal β-D-glucosyl group.

Terpenoids

This vast family of natural substances, which derives its name from the same Greek root as turpentine, is also known as the isoprene group, from the name of its basic structural unit, which is itself related to its main biosynthetic precursor, isopentenyl pyrophosphate (Chapter 8):

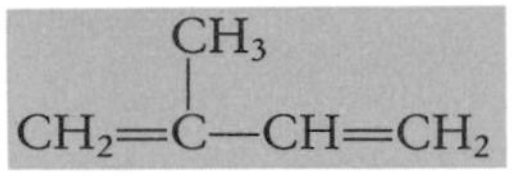

Isoprene

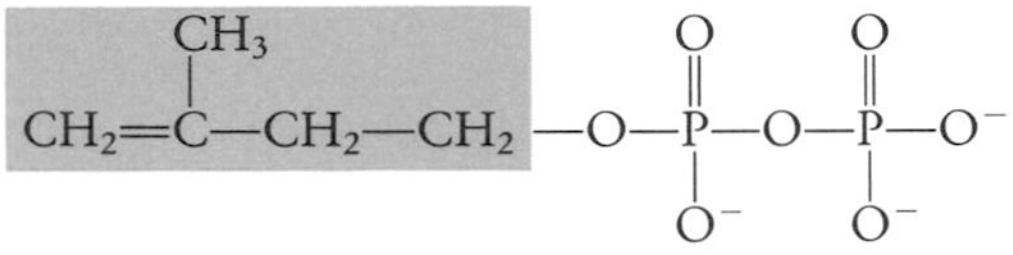

Isopentenyl pyrophosphate

An important terpene derivative is phytol, a long-chain alcohol made of four such units, which provides one of the building blocks in the formation of chlorophyll and several fat-soluble vitamins (A, E, K):

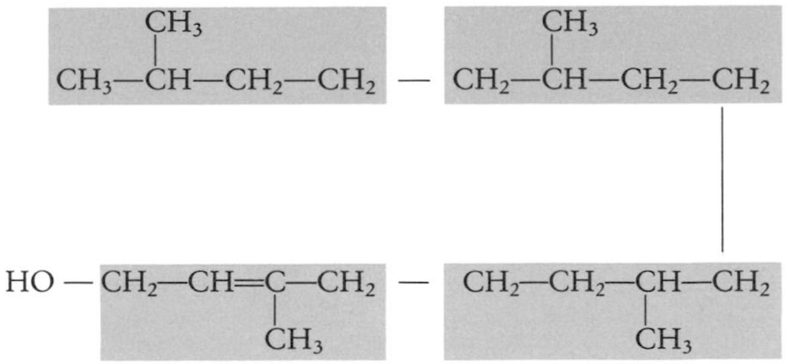

Another is squalene, a symmetrical, six-unit molecule first discovered in shark oil. Squalene has turned out to be a major biosynthetic intermediate as precursor of the steroid nucleus. The quinone electron carriers (see p. 402) and dolichol (see p. 405) are other examples of terpene derivatives, as are latex, camphor, resins, essential oils, and countless other similar plant products.

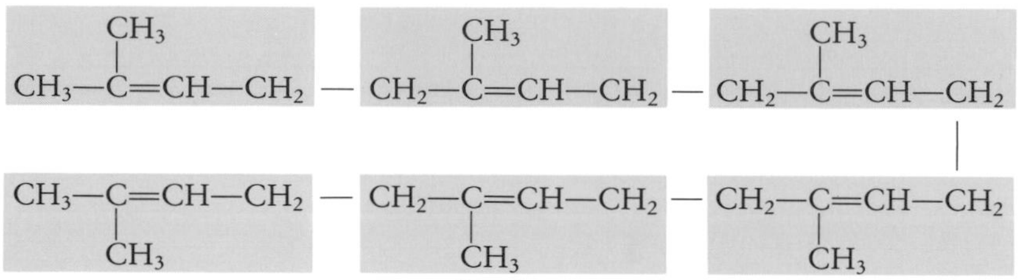

Squalene

Sterols and steroids

The mother substance of this group is cholesterol, a key constituent of the plasma membrane (Chapter 3). It is a polycyclic compound that arises from squalene by a complex process of cyclization and trimming:

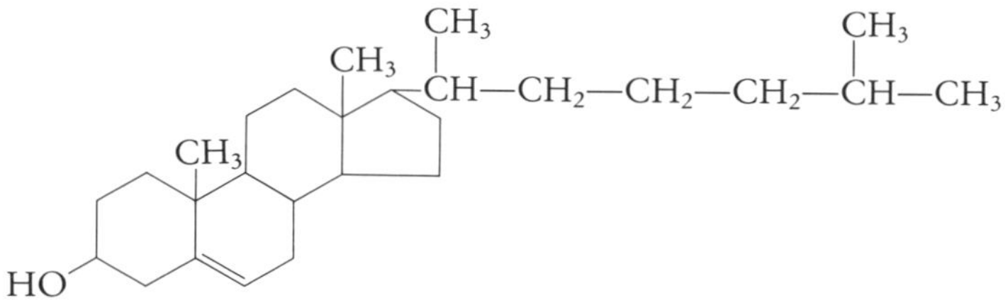

Esterification of the hydroxyl group of cholesterol by fatty acids gives cholesteryl esters, which are an essentially inert storage form. Other modifications of the cholesterol molecule lead to the bile acids, the male and female sex hormones, and cortisone and the other hormones manufactured by the adrenal cortex. Ultraviolet irradiation of cholesterol and related substances produces vitamin D.

Pyrimidines:

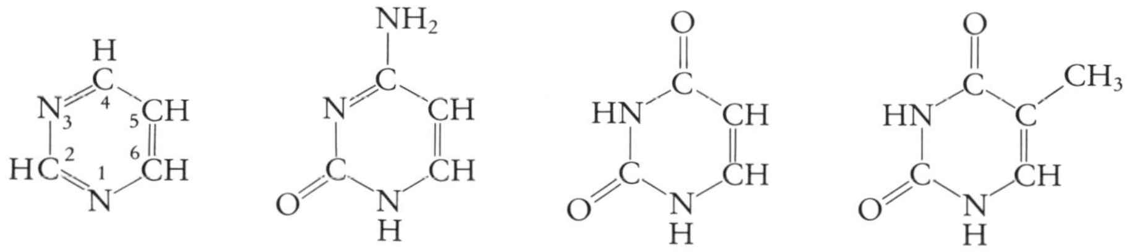

| Pyrimidine ring (Py) | Cytosine (2-keto-4-amino-Py) | Uracil (2,4-diketo-Py) | Thymine (2,4-diketo-5-methyl-Py) |

Purines:

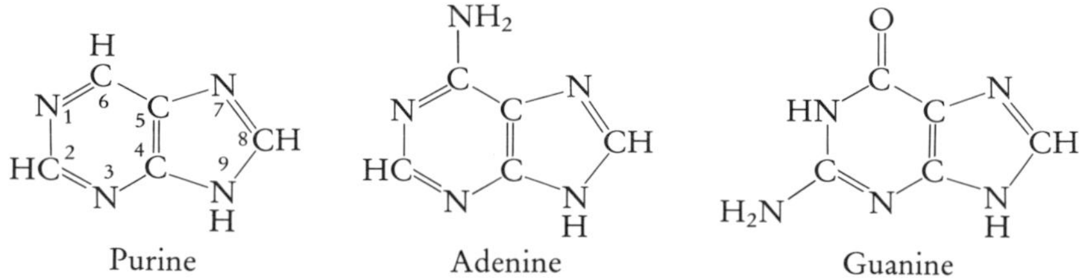

| Purine ring (Pu) | Adenine (6-amino-Pu) | Guanine (2-amino-6-keto-Pu) |

Nucleic Acids

Bases

As explained in Chapters 8, 15, and 16, nucleic acids are made of three kinds of building blocks: bases, pentoses (already seen on p. 387), and phosphoric acid. The bases are derived either from the pyrimidine or from the purine ring, as shown on the facing page.

Nucleosides

Heteroglycosides in which D-ribose or D-deoxyribose is linked to a purine or pyrimidine base are called nucleosides. The linkage is of the β-glycosidic type, with nitrogen 1 of the pyrimidine or nitrogen 9 of the purine. To avoid confusion, the atom numbers of the pentose are marked with a prime:

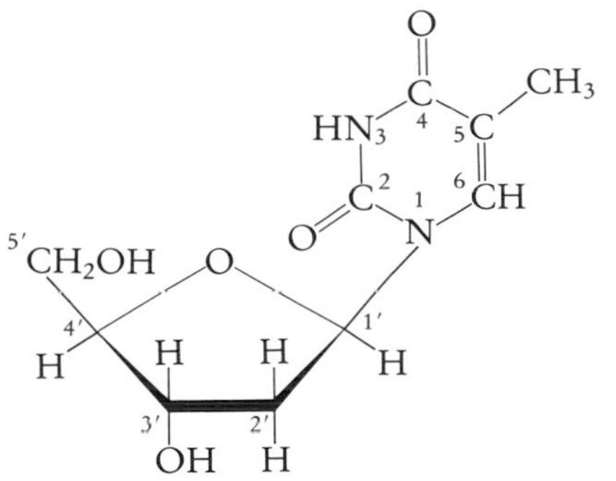

A pyrimidine deoxynucleoside
Deoxythymidine

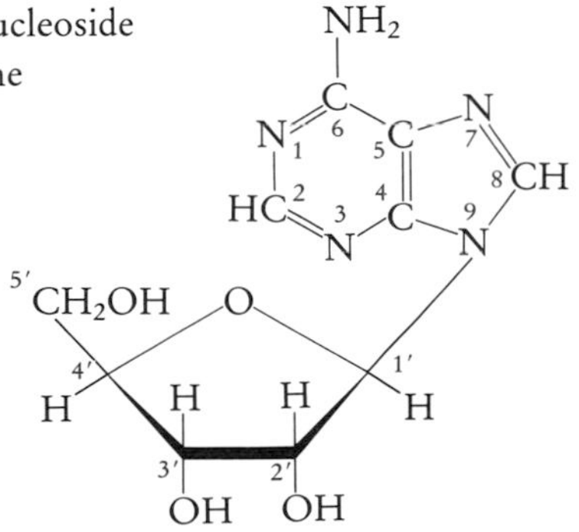

A purine nucleoside
Adenosine

The nomenclature and symbols used to denote the natural nucleosides and their derivatives are given in Chapter 8 (p. 127).

Nucleotides

Mononucleotides arise from nucleosides by phosphorylation of one of the hydroxyl groups of the pentoses, most often that in position 5′:

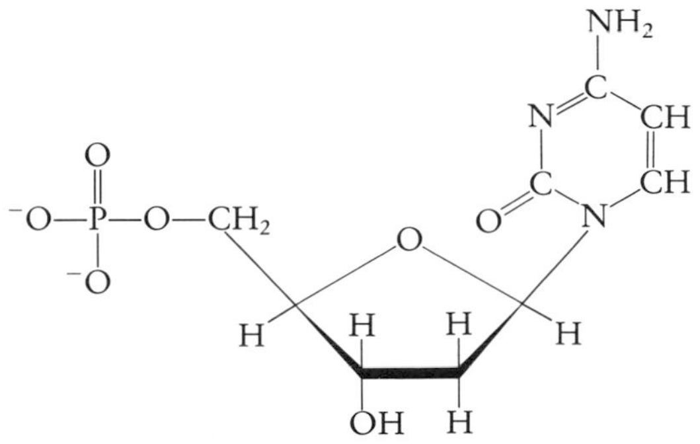

A pyrimidine 5′-mononucleotide
Deoxycytidine monophosphate (dCMP)
or deoxycytidylic acid

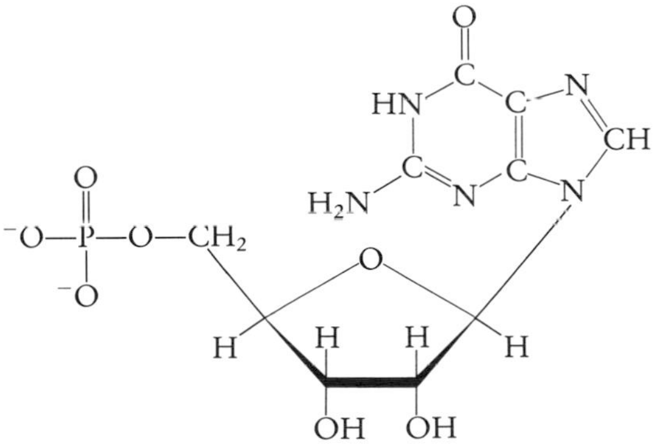

A purine 5′-mononucleotide
Guanosine monophosphate (GMP)
or guanylic acid

The phosphoryl group of 5′-mononucleotides can form several types of combinations of major biological importance. As seen in Chapter 8, it can bind one or two additional phosphoryl groups to form the various nucleoside diphosphates and triphosphates. It can also join by a second phosphoester bond with another hydroxyl of the pentose to which it is attached. Cyclic nucleotides arise in this way—for instance, cyclic AMP (Chapters 8 and 13):

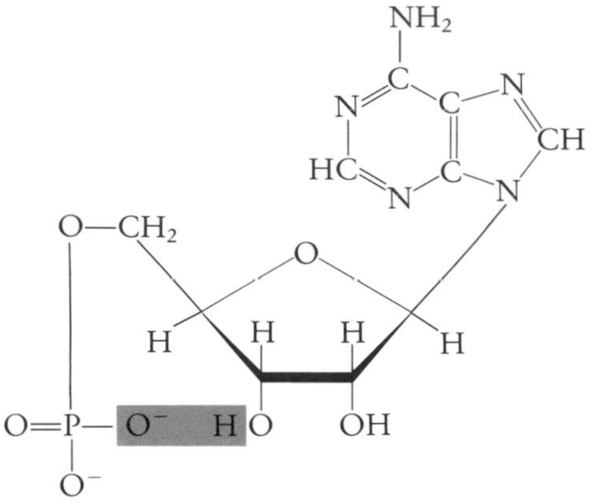

Adenosine monophosphate (AMP)

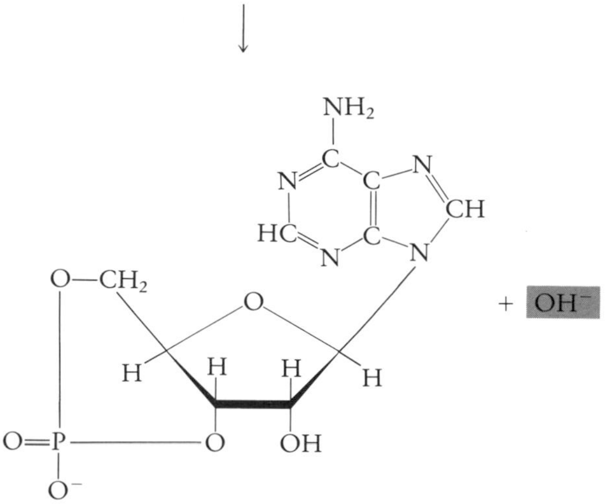

3′,5′-Cyclic AMP

In a third type of linkage, the phosphoryl group of a 5′-mononucleotide binds by a pyrophosphate bond to the phosphoryl group of another mononucleotide or analogous substance. Several important coenzymes arise in this manner (see pp. 397–399 and p. 402).

Finally, the phosphoryl group of 5′-mononucleotides can join with the 3′ hydroxyl group of another mononucleotide. Such a dinucleotide can lengthen by the similar attachment of another mononucleotide unit to its free 3′-hydroxyl group to give a trinucleotide. Repetition of the process leads to oligonucleotides and polynucleotides.

Nucleic Acids

The polynucleotide chains of nucleic acids have the following structure, in which the free valences in position 2′ are occupied either by OH (RNAs) or by H (DNAs):

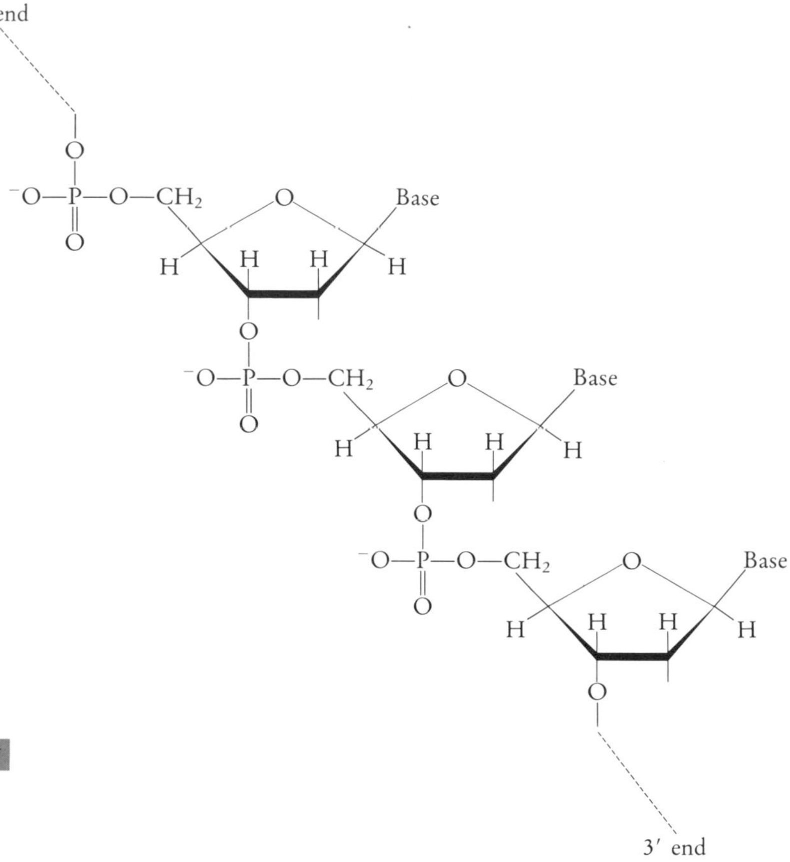

Or, in simplified form (N stands for nucleoside, p for phosphoryl):

5′ end . . . . pNpNpNpN . . . . 3′ end

The 5′ end of such chains usually bears a phosphoryl group. It may also, as in newly made RNAs (Chapter 16), bear a triphosphoryl group or, as in most eukaryotic mRNAs, a 7-methyl-GTP cap (see p. 397). In a number of cases, the terminal 5′-phosphoryl and 3′-hydroxyl groups of a chain join to produce a cyclic molecule with no free ends.

Modified Nucleic Acids

After their formation, nucleic acids may undergo a variety of chemical modifications, mostly of the bases—for example:

methylation (of adenine, guanine, cytosine, and, in some RNAs, the 2′ hydroxyl group of ribose); hydroxymethylation (of cytosine in certain phage DNAs); glycosylation of such a hydroxymethyl group (also in phages); deamination (of adenine, to give 6-ketopurine, named hypoxanthine, whose nucleoside is inosine, designated I); reduction (of the double bond between carbons 5 and 6 of uridine, to give dihydrouridine, or DHU); and transglycosylation (in uridine from nitrogen-1 to carbon-5, to give pseudouridine, designated ψ). Transfer RNAs (Chapter 15) are particularly rich in chemically modified bases. Typical examples of modified structures are:

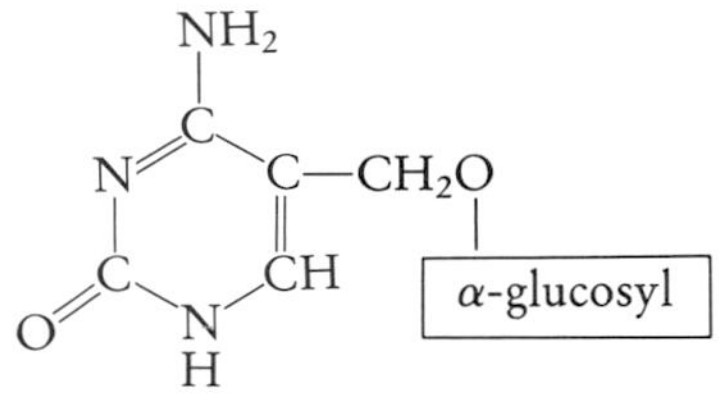

α-Glucosyl-hydroxymethylcytosine

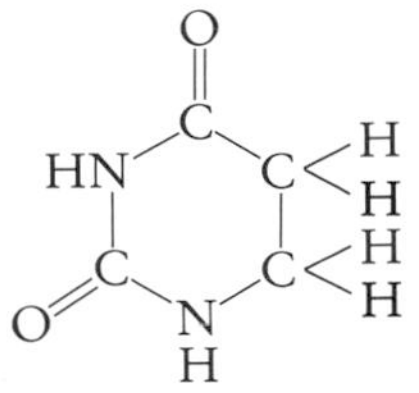

Dihydrouracil

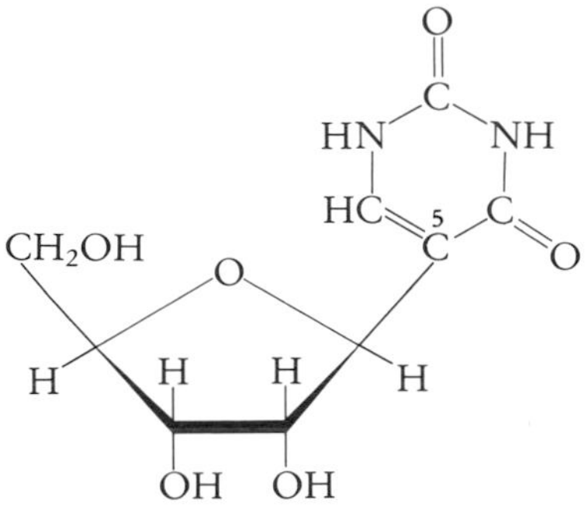

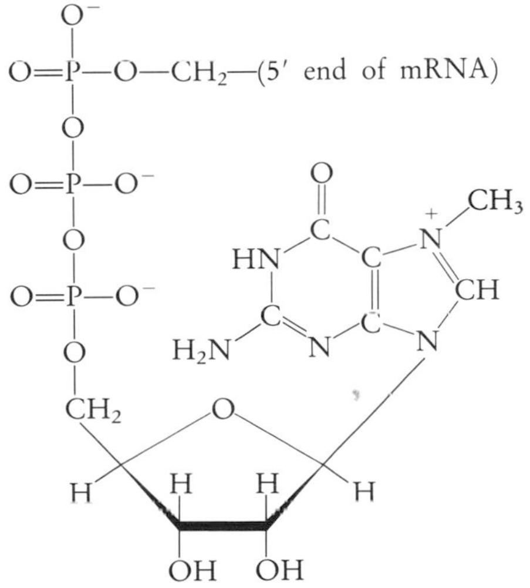

7-Methyl-GTP cap of mRNA

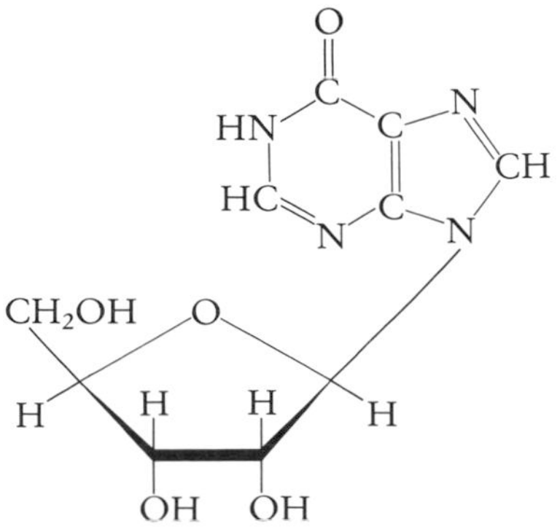

Inosine

Pseudouridine
ψ

Electron Carriers

Nicotinamide Coenzymes

Nicotinamide adenine dinucleotide (NAD), the most important coenzyme in electron-transfer reactions (Chapters 7, 9, 10, and 14) is a dinucleotide in which a molecule of AMP is linked by a pyrophosphate bond to a molecule of nicotinamide mononucleotide (NMN), a ribonucleotide in which the position of the base is occupied by nicotinamide (vitamin PP):

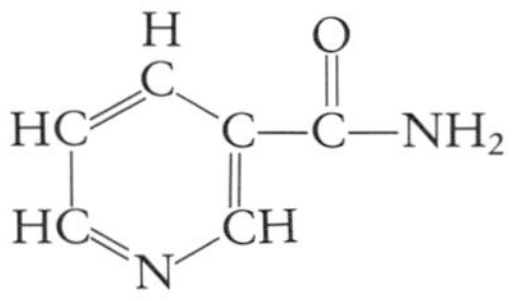

This molecule provides the electron-acceptor part of the coenzyme. Nicotinamide adenine dinucleotide phosphate (NADP), which plays a special role in photochemical and other biosynthetic reductions (Chapter 10), is NAD with an additional phosphoryl group in position 2′ of the AMP moiety:

| The Building Blocks of Living Cells

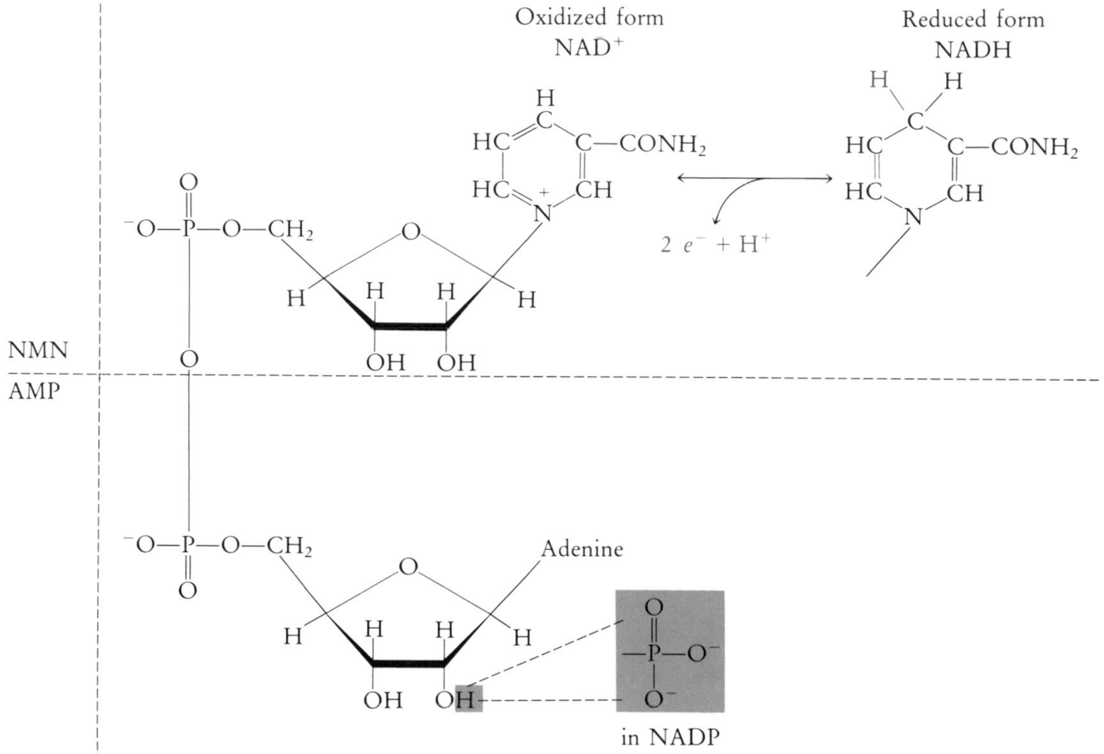

Oxidized form
NAD$^+$

Reduced form
NADH

$$2\ e^- + H^+$$

NMN
AMP

Adenine

in NADP
Nicotinamide adenine dinucleotide (phosphate), NAD(P)

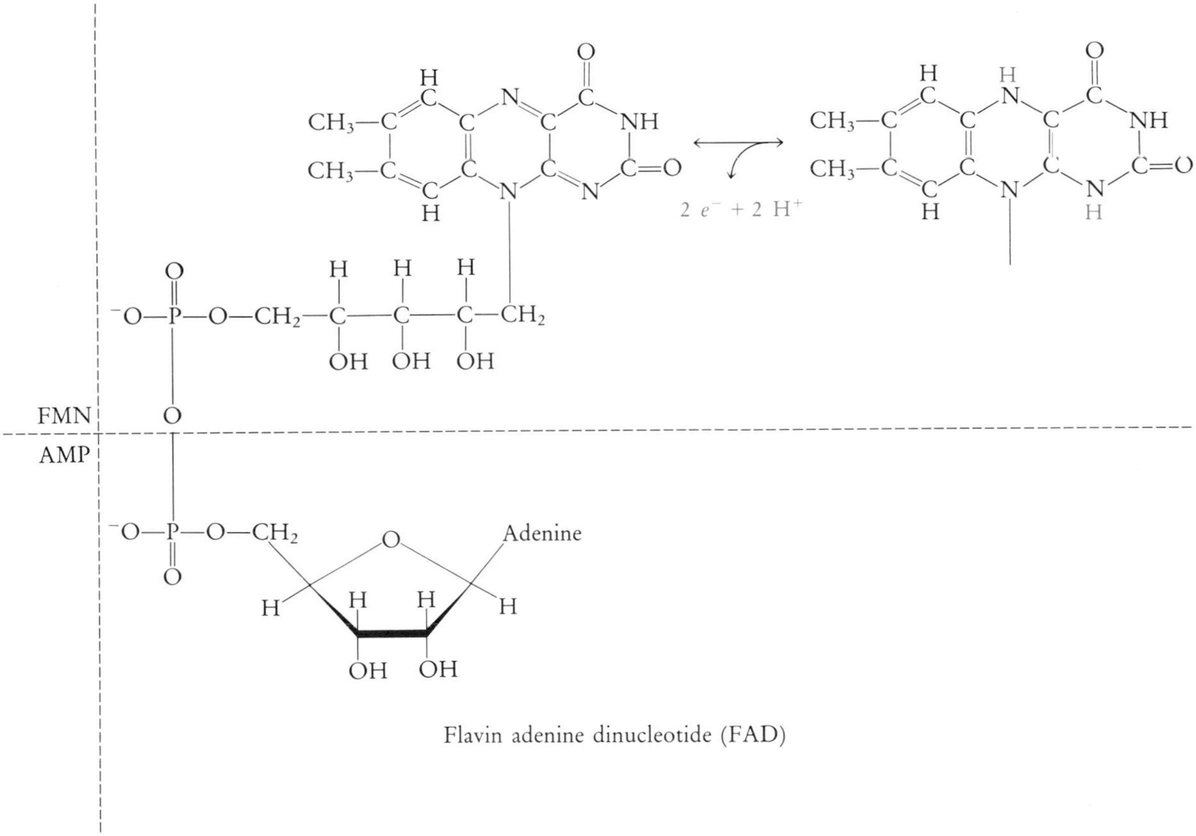

$$2\ e^- + 2\ H^+$$

Flavin adenine dinucleotide (FAD)

Flavin Coenzymes

The flavin coenzymes also participate in a number of electron-transfer reactions (Chapters 9 and 10). They include flavin mononucleotide (FMN) and flavin adenine dinucleotide (FAD), which consists of FMN linked to AMP by a pyrophosphate bond, as are the two mononucleotide parts of NAD. The nucleoside part of FMN is really a pseudonucleoside. Known as riboflavin (vitamin B_2), it consists of the pentitol ribitol, the reduction product of ribose, and of a derivative of the polycyclic isoalloxazine ring, which is the reactive part of the coenzymes.

Porphyrin Cofactors

The heme groups of hemoglobins, cytochromes, and other hemoproteins and the chlorophylls of photosynthetic organisms all derive from the porphin, or tetrapyrrole, ring—a planar, polycyclic molecule made of four pyrrole rings linked by methene (—CH=) bridges. Porphyrins are derived from the porphin ring by various substitutions on carbons 1 through 8 and sometimes on some of the methene bridges.

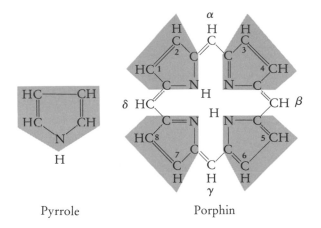

Pyrrole   Porphin

Hemes are iron-porphyrin complexes in which the two central hydrogens are replaced by an iron atom. The most important such molecule is protoheme IX, derived from protoporphyrin IX, or 1,3,5,8-tetramethyl-2,4-divinyl-6,7-dipropionic acid porphin:

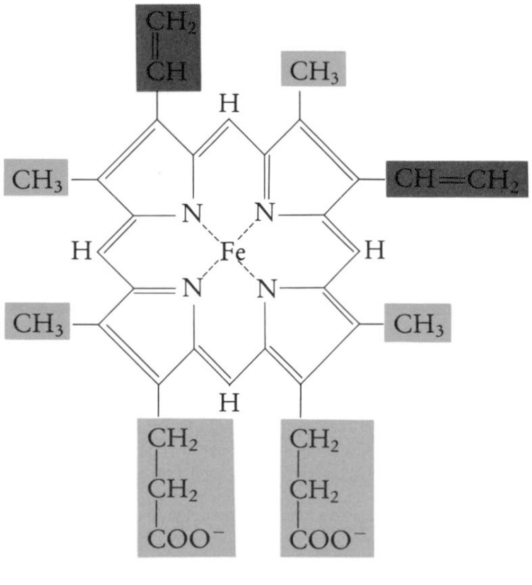

This molecule associates with a variety of proteins to give rise to hemoproteins. In hemoglobins, the central iron is in the ferrous (Fe^{2+}) form and serves as binding site for oxygen. It is in the ferric (Fe^{3+}) form in peroxidases and catalase and binds hydrogen peroxide, the electron acceptor used by these enzymes. In cytochromes, the iron alternates between the two forms to

serve in electron transport. These three possibilities are shown schematically as follows with the porphyrin ring cut perpendicularly to its plane through two of the four central nitrogens:

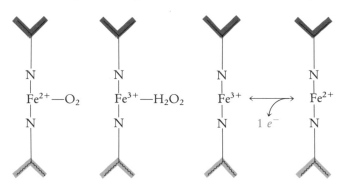

| O_2 complex of hemoglobins | H_2O_2 complex of peroxidases and catalase | Electron transport by cytochromes |

Cytochrome oxidase is a complex molecule made of two distinct cytochromes (their heme component is different from protoheme IX) and containing copper. It catalyzes a concerted transfer of four electrons to a bound molecule of oxygen:

$$O_2 + 4\ e^- + 4\ H^+ \longrightarrow 2\ H_2O$$

In chlorophylls, the central position of the porphyrin is occupied by magnesium, and the substituents include a molecule of phytol (p. 393).

Metalloproteins

Iron also participates in electron transport in nonheme form as protein-bound iron-sulfur complexes. These consist of a single iron atom, of two irons and two sulfides (Fe_2S_2), or of four irons and four sulfides (Fe_4S_4), in each case cradled by a cluster of four cysteine residues belonging to the protein. The schematic structure of an Fe_2S_2 complex is shown on the facing page.

The iron in iron-sulfur proteins oscillates between the ferrous and ferric states, as in hemoproteins. Some of the iron-sulfur proteins, called ferredoxins, accept electrons at a particularly high level of energy (low redox potential) and participate in this capacity in the photoreduction of NADP (Chapter 10) and the formation of molecular hydrogen (Chapter 11). The respiratory chain of mitochondria also includes iron-sulfur proteins.

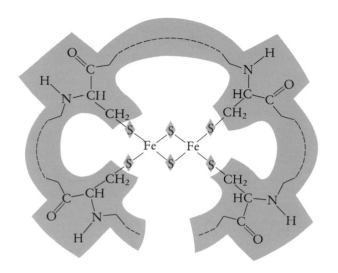

Another important thiol is lipoic acid, which bears two sulfhydryl groups on an eight-carbon fatty-acid skeleton. Its carboxyl group is normally attached by an amide bond to the free amino group of a lysine residue belonging to the enzyme for which it serves as coenzyme. Upon oxidation, lipoic acid forms an intramolecular disulfide bridge:

$$CH_2-CH_2-CH-CH_2-CH_2-CH_2-CH_2-\overset{\overset{\displaystyle O}{\|}}{C}-NH$$
$$\underset{SH}{|} \qquad \underset{SH}{|} \qquad\qquad\qquad\qquad Enzyme$$

$$2\ e^- + 2\ H^+$$

$$CH_2-CH_2-CH-CH_2-CH_2-CH_2-CH_2-\overset{\overset{\displaystyle O}{\|}}{C}-NH$$
$$\underset{S}{|}\text{————}\underset{S}{|} \qquad\qquad\qquad\qquad Enzyme$$

In its reduced form, lipoic acid also functions as an acyl carrier, like coenzyme A (see pp. 402–403). Thanks to this dual ability to participate in both electron and group transfer, lipoic acid plays a key role in several important substrate-level phosphorylation reactions (Chapters 7 and 8). Because of the covalent attachment of lipoic acid to the enzyme, its transporting capacity is limited to the distance that can be covered by its flexible arm. We will encounter a similar situation with biotin. (see p. 403).

Other metals also play a role in some electron-transfer reactions. The most important ones are copper and manganese, which are found in a number of oxidizing enzymes.

Thiols

The reaction whereby two molecules of cysteine combine oxidatively by a disulfide bridge can play a role in electron transport:

$$2\ R-SH \rightleftharpoons R-S-S-R + 2\ e^- + 2\ H^+$$

Particularly important in this respect is the tripeptide glutathione (GSH), which dimerizes oxidatively to GSSG (GSH is γ-glutamylcysteinylglycine):

$$2\ GSH \rightleftharpoons GSSG + 2\ e^- + 2\ H^+$$

Quinones

Quinones, which can be reduced to the corresponding diphenols, act in electron transfer in the respiratory chains of both mitochondria and chloroplasts. They are generally derived from p-benzoquinone and bear various substituents, including a long terpenoid chain. The mitochondrial coenzyme Q_{10}, or ubiquinone 50, has the following structure:

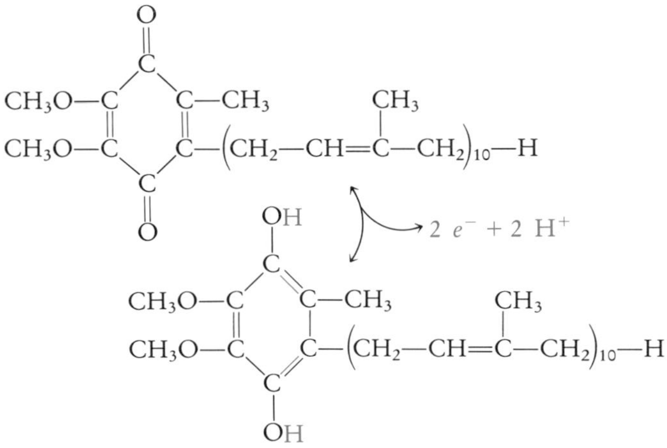

The number 10 in Q_{10} identifies the number of isoprene units in the side chain, whereas the number 50 affixed to ubiquinone refers to the number of carbon atoms of the side chain.

Group Carriers

Phosphagens

The phosphagens—literally phosphate generators—are phosphorylated substances of high group potential that serve as reservoirs of high-energy phosphoryl groups for the rapid regeneration of ATP from ADP in muscle tissue (Chapter 8). They have the distinction that the phosphoryl group is attached to a nitrogen atom belonging to a guanidine group. In many invertebrates, the phosphoryl carrier is the amino acid arginine. It is creatine, a derivative of the amino acid glycine, in vertebrates.

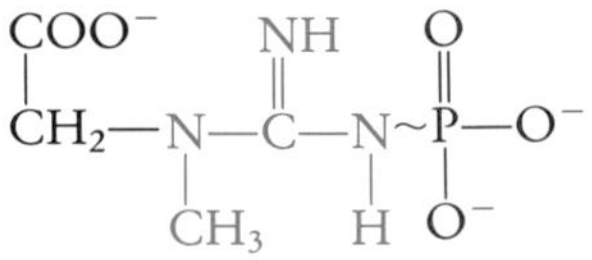

Creatine phosphate 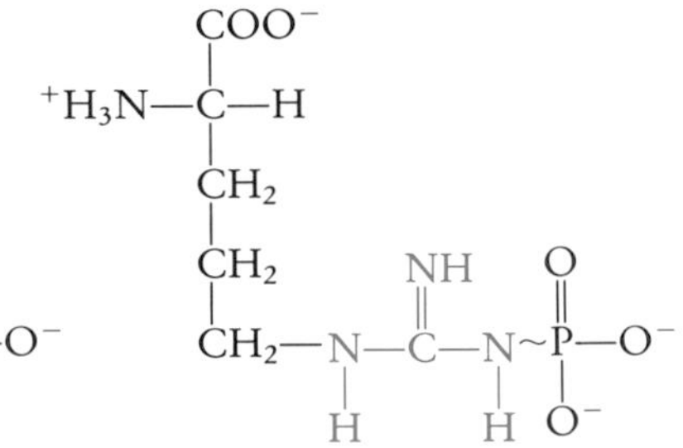   Arginine phosphate

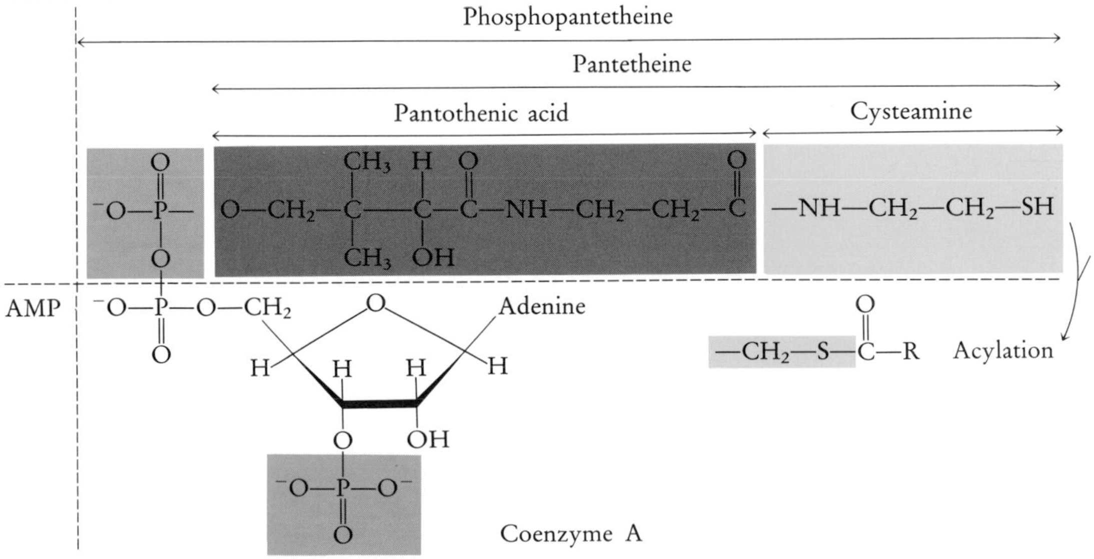

Coenzyme A

Coenzyme A

This central carrier of acyl groups (Chapter 8) consists of a molecule of 3′-phospho-AMP linked by a pyrophosphate linkage to phosphopantetheine, which itself is formed through the association, by an amide bond, of cysteamine (the product of decarboxylation of the amino acid cysteine) with pantothenic acid, or vitamin F. The thiol group of cysteamine serves as carrier for the acyl group. The structure of coenzyme A is shown on the facing page.

In acyl-carrier protein (ACP, Chapter 8), the phosphopantetheine part of coenzyme A is attached to the protein without the participation of AMP.

Carnitine

Carnitine (Chapter 13) is a fatty acyl carrier that serves in the transfer of coenzyme A-linked fatty acyl groups across the inner mitochondrial membrane:

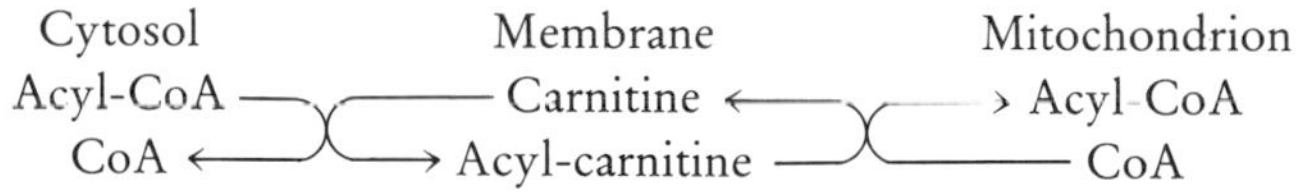

Cytosol	Membrane	Mitochondrion
Acyl-CoA ———	Carnitine ←	→ Acyl-CoA
CoA ←	→ Acyl-carnitine ———	CoA

Carnitine is a derivative of choline, already encountered as a constituent of phospholipids, and of acetic acid. These two moieties are attached by a carbon–carbon bond, not by an ester bond as in acetylcholine. The hydroxyl group of the choline part serves as acyl carrier:

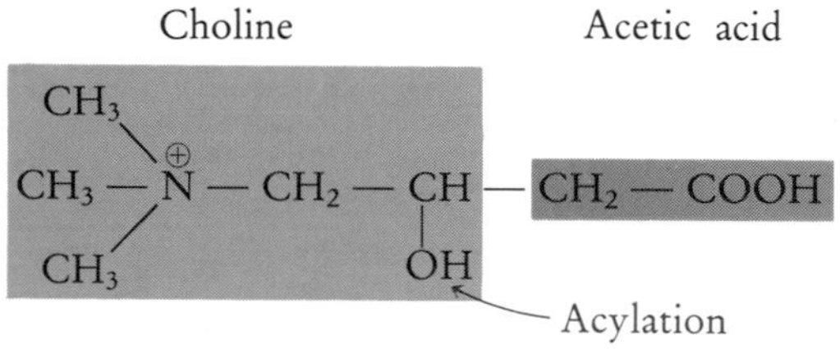

Choline   Acetic acid

Acylation

Biotin

Biotin (vitamin H), the carboxyl carrier in carboxylation reactions (Chapter 8), is a saturated sulfur-containing molecule fitted with a five-carbon acid side chain, which is covalently attached to the transferase molecule and serves as a flexible arm in the transport of the activated carboxyl group to its final acceptor. (Lipoic acid, considered on p. 401, presented us with a similar situation.) The carboxyl group is attached to one of the nitrogens of the ring:

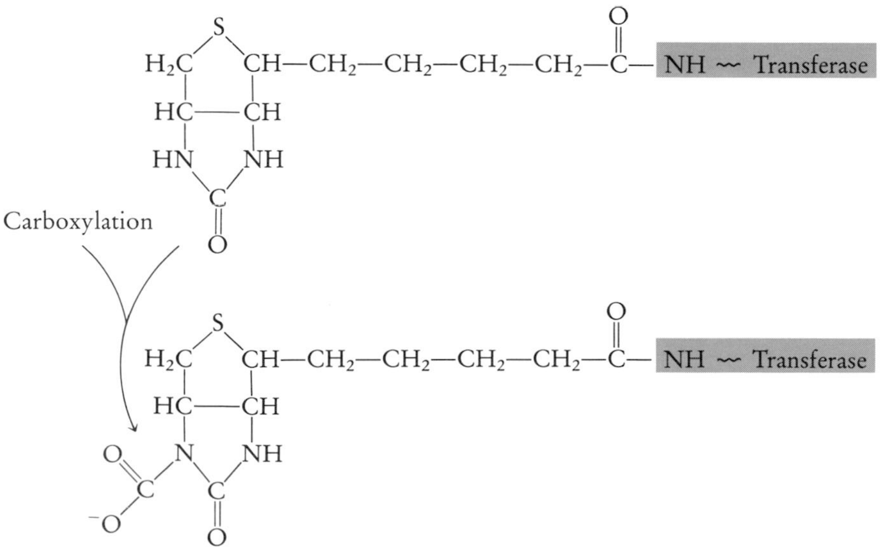

Carboxylation

403 | The Building Blocks of Living Cells

Folic acid

Three components are assembled in the formation of this vitamin of the B group: a derivative of the pterine ring, *p*-aminobenzoic acid, and glutamic acid:

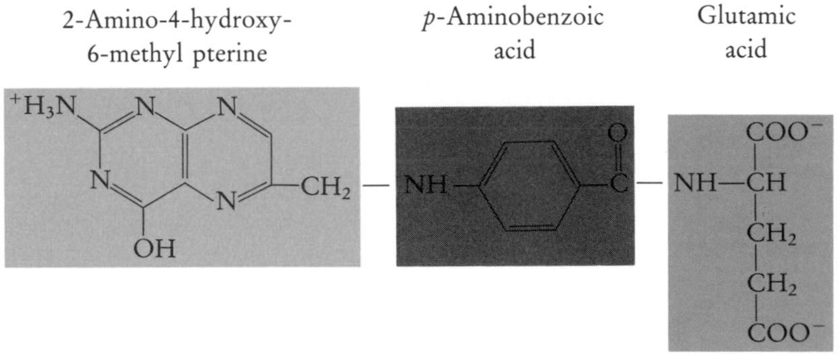

2-Amino-4-hydroxy-6-methyl pterine	*p*-Aminobenzoic acid	Glutamic acid

As pointed out in Chapter 8, tetrahydrofolic acid (THF), the product of reduction of folic acid, serves in the transport of the formyl group and of a number of its derivatives. Following is the structure of formyl-THF:

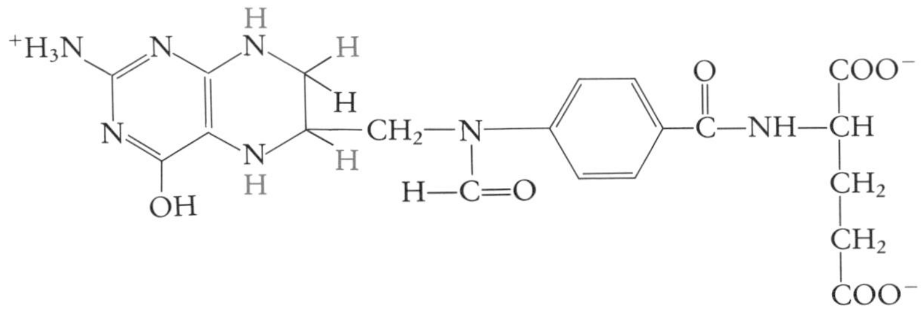

Dolichyl phosphate

As mentioned in Chapters 6, 8, and 13, dolichyl phosphate plays an important role as glycosyl carrier in the assembly of oligosaccharide side chains of glycoproteins and glycolipids in the ER. It is a very long terpenoid alcohol containing about twenty isoprene units. This highly hydrophobic chain serves to anchor the molecule firmly in the lipid bilayer of the ER membrane. Phosphorylation of the alcohol group provides the molecule with a hydrophilic head on which the transported glycosyl groups are attached by a glycosyl phosphate bond:

$$
\underset{\substack{\\ \text{CH}_3}}{\text{CH}_3-\text{C}=\text{CH}-\text{CH}_2} - \left[\underset{\substack{\\ \text{CH}_3}}{\text{CH}_2-\text{C}=\text{CH}-\text{CH}_2} \right]_n - \underset{\substack{\\ \text{CH}_3}}{\text{CH}_2-\text{CH}-\text{CH}_2-\text{CH}_2} - \text{O}-\overset{\text{O}}{\underset{\text{O}^-}{\overset{\|}{\text{P}}}}-\text{O}^-
$$

$$(n = 15 \text{ to } 19)$$

[Glycosyl$^{\oplus}$]
or
[Glycosyl-phosphoryl$^{\oplus}$]

In some reactions, dolichyl phosphate carries a glycosyl phosphate molecule linked by a pyrophosphate bond.

Thiamine

First of the vitamins to be discovered, thiamine, or vitamin B_1, was singled out in Chapter 8 as one of the rare recipients of the pyrophosphoryl group. The product of this reaction, thiamine pyrophosphate (TPP), serves as coenzyme in a number of decarboxylation reactions, such as the decarboxylation of pyruvic acid to acetaldehyde in alcoholic fermentation (Chapter 7). TPP is highly unusual among group carriers in that it serves to transport a negatively charged group, the carbanion group $R-CO^-$. Thiamine is formed by the association of a pyrimidine derivative with a thiazol derivative:

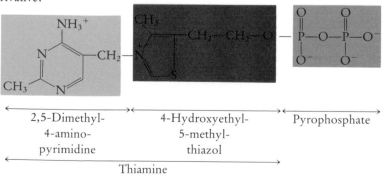

| 2,5-Dimethyl-4-amino-pyrimidine | 4-Hydroxyethyl-5-methyl-thiazol | Pyrophosphate |

Thiamine

Pyridoxal phosphate

Derived from pyridoxine, or vitamin B_6, pyridoxal phosphate is the coenzyme of transamination reactions. These group-transfer reactions do not conform to the general scheme of group transfer given in Chapter 8. Instead of being exchanged for a proton, the transferred amino group is exchanged for a ketonic oxygen. As a rule, the partners of the reaction are an amino acid and an α-keto acid:

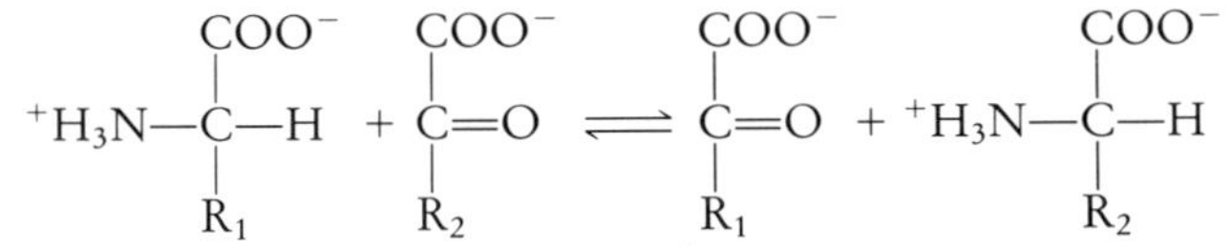

$$
\begin{array}{ccccc}
\text{COO}^- & \text{COO}^- & & \text{COO}^- & \text{COO}^- \\
| & | & & | & | \\
{}^+\text{H}_3\text{N}{-}\text{C}{-}\text{H} & +\ \text{C}{=}\text{O} & \rightleftharpoons & \text{C}{=}\text{O} & +\ {}^+\text{H}_3\text{N}{-}\text{C}{-}\text{H} \\
| & | & & | & | \\
\text{R}_1 & \text{R}_2 & & \text{R}_1 & \text{R}_2
\end{array}
$$

Pyridoxine is a pyridine derivative (like nicotinamide). It is 2-methyl-3-hydroxy-4,5-dihydroxymethylpyridine. In its functional form, it is phosphorylated in position 5 and either oxidized (pyridoxal phosphate) or aminated (pyridoxamine phosphate) in position 4:

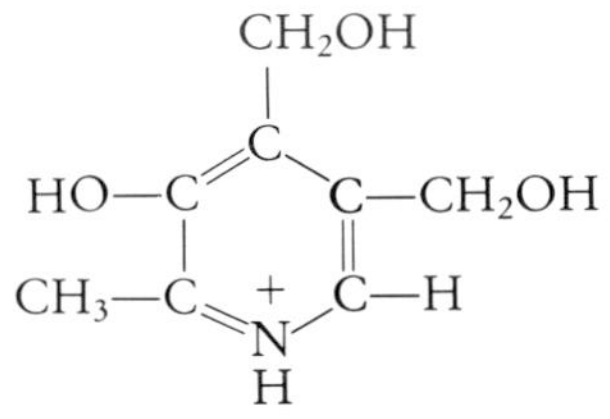

Pyridoxine 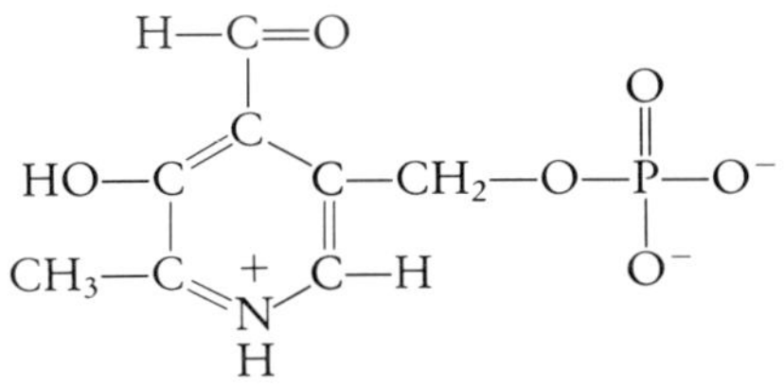   Pyridoxal phosphate 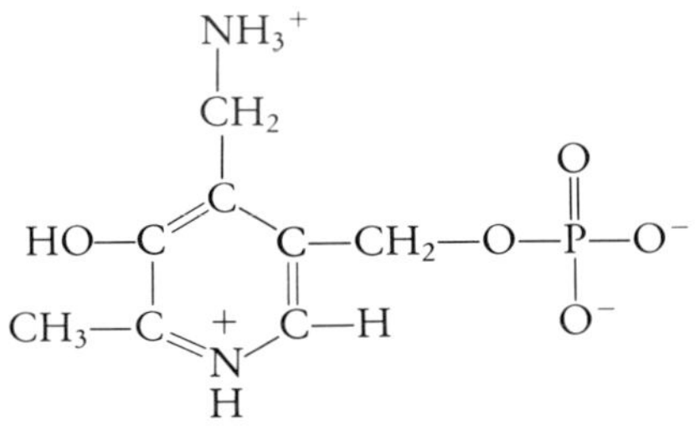   Pyridoxamine phosphate

The involvement of the coenzyme in the transaction is due to its ability to form Schiff bases reversibly with amino acids (as pyridoxal) and with α-keto acids (as pyridoxamine). Isomerization of the Schiff base allows the exchange of groups between coenzyme and substrate in the manner illustrated below. As indicated, an amino group is exchanged for a ketonic oxygen at each turn of the cycle, while the coenzyme oscillates between its pyridoxal and pyridoxamine forms. Every step in this sequence is, of course, reversible. One-way arrows have been used to show a complete reaction cycle:

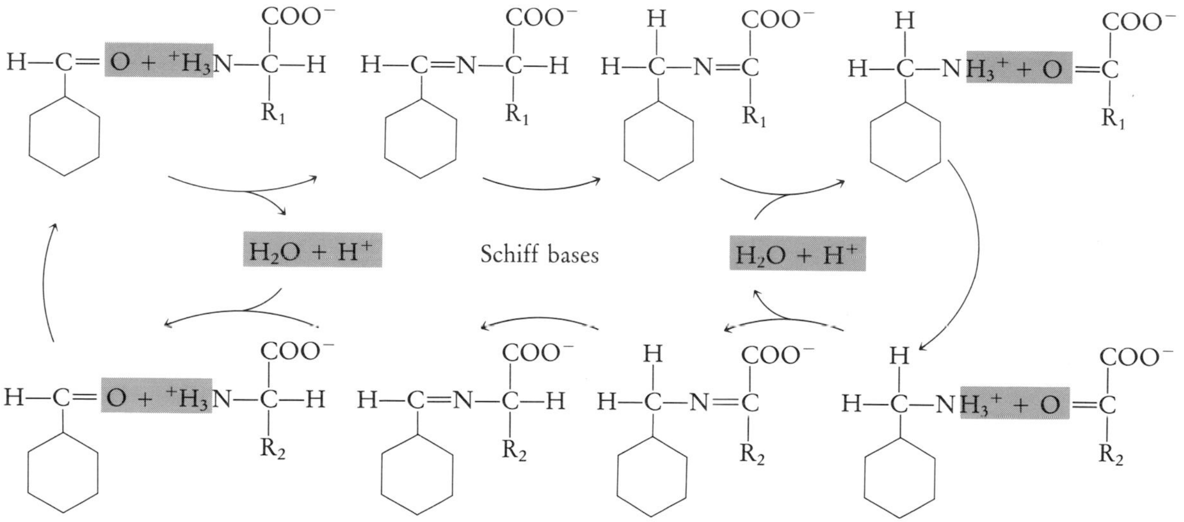

APPENDIX 2 | Principles of Bioenergetics

Free Energy, the Source of Work

In terms of everyday experience, energetics is very simple: when we perform work, we do so at the expense of something that enables us to do work. That something we call energy. Scientists, when they want to be rigorous, call it *free energy*, to take care of the fact that the conversion of certain forms of energy—heat, for example—into work is subject to constraints.

As we all know from having felt exhausted after physical exertion, the more work we perform, the more (free) energy we spend. We also know, although these concepts are less intuitive, that there are different kinds of work—mechanical, electric, chemical—and that they are interconvertible under certain conditions; that is, one type of work can supply free energy for another with the help of an appropriate *transducer.* The electric generator in our automobile is an example of a mechanoelectric transducer. The main practical function of *energetics* (originally named *thermodynamics* because it was developed to explain and assess the performance of steam or heat engines) is to put these familiar notions into quantitative terms to allow accurate bookkeeping of the free-energy gains and losses for every kind of work and, thus, for every kind of conversion. *Bioenergetics* does this for living organisms, which, it should be stressed, obey exactly the same laws as do inanimate objects. But living organisms carry out a remarkable set of energy conversions with the help of unique transducers.

All of Nature's Streets Are One Way

Consider the following statements: "Newton's apple could not, under its own power, jump back to the branch from which it fell" or "Nobody has ever seen, or expects to see, sugar gathering back into a lump in a cup of coffee." "Unexceptionable," you will say, or perhaps even "Self-evident! Do you take me for a moron?" Yet, if I write:

$$\Delta G \leq 0 \qquad (1)$$

or state that "The free energy of an isothermic and isobaric closed system that can exchange only heat with its surroundings cannot increase spontaneously," you may well ask with equal furor whether I take you for Einstein.

Actually, both types of statements say the same thing: Some events are possible; others are not. The more obscure statement simply uses the formalism that scientists had to devise to put the obvious into rigorous, quantitative terms of universal significance. It expresses what is known as the second law of thermodynamics. The first law is the law of *conservation of energy* (now combined with the law of *conservation of matter*, since Einstein's demonstration that matter and energy are interconvertible); it is easy to grasp. The second law, sometimes called the law of *degradation of energy*, is more abstruse. Much of energetics is concerned with its applications.

Free energy has already been defined as the ability to perform work. The italic letter G is the symbol chosen to represent this quantity. The capital Greek delta, Δ, stands for *difference*—between a final and an initial state. Therefore, ΔG represents the *change in free energy*—positive if there is a gain, negative if there is a loss—that is associated with a given event or transformation. When we write, as in equation 1, that ΔG is smaller than, or equal to, zero—that is, either negative or null—we actually state an interdiction: ΔG cannot be positive; G cannot increase spontaneously. Which, when you come to think of it, is the reason why Newton's apple does not levitate: it would thereby increase its G content, its ability to perform work by falling; and Newton would get something for nothing. Similarly, although this may be less obvious, dissolved sugar does not spontaneously reform into a lump because this would, at no cost, reinstate its initial ability to perform work, of an osmotic type in this case, by dissolving.

This, one can readily appreciate once it is pointed out, is an absolutely general observation. For every natural phenomenon, there is an authorized and a forbidden direction. Stones roll downhill, not uphill; heat flows from hot to cold, not in the reverse direction; molecules diffuse from regions where they are more abundant to regions where they are less; Nature abhors a vacuum; the list is endless. Energetics gives us the universal key to unidirectionality by telling us that the one-way sign always points in the direction of decreasing G. This direction is defined

The difference in ability to perform work between final state and initial state is indicated by ΔG.

1. Endergonic transformation: G increases, $\Delta G_1 > 0$.

2. Exergonic transformation: G decreases, $\Delta G_2 < 0$.

Only exergonic transformations can take place spontaneously. However, an endergonic transformation can be made to occur by coupling to an exergonic transformation, provided $\Delta G_1 + \Delta G_2 < 0$.

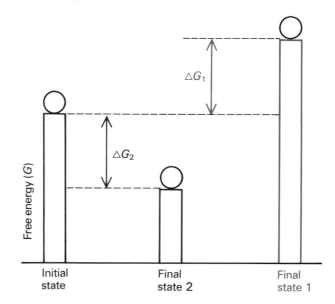

by the term *exergonic*, as opposed to *endergonic*, when G increases. The special case in which ΔG equals zero characterizes the state in which no change can occur, whether in one direction or in the other, because G is at a minimum. This state of authentic standstill is known as *stable* or *thermodynamic equilibrium*.

So far, so good. But we must watch out. Falling apples can be lifted back to their starting position; sugar can be crystallized; stones can be pushed uphill; heat can be forced out of a refrigerator; vacuum can be created. In other words, prohibited phenomena can be made to take place. The telling word, here, is "made." Endergonic events never occur spontaneously, but they can be made to occur, which, of course, is what work is all about. When we work, or have a machine work for us, it is always to accomplish something that would not happen on its own, to back up one of Nature's one-way streets. This is not against the law, provided we spend at least as many G units in doing the work as we gain from it. Let ΔG_1 be the free-energy change (positive) of the endergonic event that has been accomplished and ΔG_2 the free-energy change (negative) of the exergonic event that drives the former. Then, if ΔG_2 is equal to, or larger than, ΔG_1 in absolute value, we may write:

$$\Delta G_1 + \Delta G_2 \leq 0 \qquad (2)$$

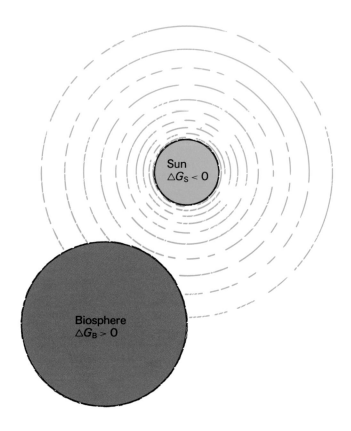

Life is supported by the sun; it does not violate the second law because $\triangle G_B + \triangle G_S < 0$.

For the combined events, the second law is obeyed; Nature's ways have not been violated. There is a net loss of free energy—or, at least, no net gain—although part of the system has gained free energy. Thus, free-energy gains can take place, provided they are associated with free-energy losses of at least the same magnitude. An essential condition of success in such an operation is *coupling* between the two events. Machines, including the "bio-engines" that we see at work in the living cell, are no more than coupling devices of this sort. The *efficiency* of a machine is given by the ratio of free energy gained to free energy lost. Expressed in percentage:

$$\% \text{ efficiency} = \frac{\Delta G_1}{-\Delta G_2} \times 100 \qquad (3)$$

As a rule, the efficiency is lower than its maximal permissible value of 100 per cent, and a fraction of the free energy expended is lost as heat. We will come back to this point. Note that the loss can be very small. Especially in the living cell, efficiency values approaching 100 per cent are not uncommon, which means that the coupled systems concerned operate near thermodynamic equilibrium.

Of Systems and Their Boundaries

With energetics, we can find out what allows any part of the world to run as it does. But, as illustrated by the examples that we have just seen, we must make sure of exactly what part of the world we are looking at. Should we focus on the biosphere, for example, without including the sun, we would be left with the impression of a tremendous build-up of free energy—as fuel, to say the least—characterized by an enormous positive ΔG; we might even conclude, as many did in the past, that life does not obey the second law of thermodynamics. But, if we add the even more enormous negative ΔG associated with the provision of light to photosynthetic organisms, the need to believe that life is outside the law vanishes. In order to avoid such traps and to respect the requirements of perfect rigor, the scientists who worked out the energetic code have, like lawyers, had to resort to a forbiddingly formal language, stuffed with precautionary clauses and hair-splitting definitions that have dampened many enthusiasms. The sentence that made you exclaim that you are not an Einstein belongs to that pedantic-sounding, but indispensable, phraseology.

First the word *system*. In energetics, it designates any kind of defined assembly of matter. It can be you or me, a living cell, a mitochondrion, a reaction vessel in a chemical plant, a locomotive, the earth, even the universe—anything material that can be defined by appropriate limits, even imaginary ones. A system may be open, closed, or isolated. An *open system* is one that can exchange both matter and energy with the outside. A *closed system* can exchange only energy, not matter. An *isolated system* can exchange nothing. Living organisms are open systems, traversed by fluxes of matter and energy. As such, they belong to the domain of *nonequilibrium thermodynamics*, which uses a fairly sophisticated kind of formalism. For practical purposes, it is often possible to simplify a bioenergetic problem by artificially closing the system under consideration and applying to it the laws of *equilibrium* or *classical thermodynamics*. This is what we have done.

Several additional assumptions must be made about the closed systems that we are looking at if we are to apply the second law in the simple form enunciated above. First we assume, unless explicitly stated otherwise and taken into account, that heat is the only form of energy that the systems can exchange with the outside. We assume further that this exchange is carried out at a perfectly constant temperature. Finally, we assume that the systems are kept at constant pressure (or at constant volume, but the assumption of constant pressure is clearly more appropriate for living organisms).

The assumption of *isothermicity* implies the existence around the system of a reversible heat exchanger, connected with a constant-temperature calorie reservoir of infinite capacity; so any change inside the system that would tend to raise or lower its temperature is compensated immediately, continually, and perfectly by an appropriate flow of heat out of or into the system. Heat exchangers of this sort do not exist, of course; heat flows only if there is a difference in temperature between two spots. But this does not matter. We can just think of the temperature difference as being "infinitesimally small," "negligible," or whatever. Physics—remember the "perfect gas"—needs such idealized "limit" situations to formulate manageable "laws," which can then be applied, and adapted if need be, to the real world.

The assumption of *isobaricity* (constant pressure; Greek *baros*, weight) means that, if the change affecting the system is such as to raise or lower its internal pressure, the system is free to expand or contract at constant pressure. This is an idealized situation; in practice, a little extra pressure is needed to set off such a volume change. The corresponding work performed or received by the system is taken to be automatically compensated by an equivalent flow of heat into or out of the system through the heat exchanger just mentioned. This heat is automatically included in the bookkeeping.

The Two Sources of Free Energy

Systems that perform work generally draw mostly on their own supply of *internal energy*. We can readily find out how much energy they have available for doing the work by uncoupling the transformation that supplies the energy from the work process that utilizes it—the equivalent of disengaging a clutch and letting a motor idle. The energy then appears as heat. In chemistry, it is called the *heat of reaction* and is represented by ΔH.

In agreement with the general convention followed in thermodynamics, ΔH is defined from the point of view of the sys-

tem; it represents the change in internal energy—more precisely, but less intuitively, *enthalpy* (Greek *thalpein,* to heat) for a system at constant pressure—suffered by the system in association with its transformation. If ΔH is positive, this change is a gain: energy must be supplied from the outside for the transformation to occur at constant temperature; the transformation is *endothermic*. If ΔH is negative, the system loses energy; the transformation is *exothermic*.

Should a system's capacity to work depend simply on an adequate supply of internal energy, as our personal experience of physical toil might make us believe, there would be no need to introduce the concept of free energy and to distinguish between ΔG and ΔH. But a second factor, much more elusive, may allow the system to draw additional free energy from external heat or, on the contrary, may force it to supply heat to the outside at the expense of its own internal energy. This factor confronts us with the more subtle notions of thermodynamics—namely, the nature of heat as a *degraded* form of energy and, especially, the concept of *entropy* (Greek *tropê*, turn).

Heat, or caloric energy, is the sum of the individual motion energies of all the atoms and molecules that compose a system. It is, thus, essentially kinetic in nature, as is the energy of a moving projectile. But it is random instead of directed. This is an absolutely fundamental difference, which explains why heat so often lies at the end of Nature's one-way streets. It can always be generated, but it is not readily harnessed. A moving projectile can dissipate its entire kinetic energy and convert it into heat—upon hitting an armor-plated surface, for example; nobody has ever seen a bullet picking up speed by cooling. Yet the steam engine works, and its offspring, the science of thermodynamics, was not misnamed.

Indeed, the founders of thermodynamics have solved the problem for us: it is a question of degree of randomness. There is something in the very fabric of the universe that favors randomness over regularity, chaos over order, disorganization over structure. This is so obvious, so much a matter of everyday experience, that it hardly needs demonstrating. Actually, it is very easy to demonstrate, at least in principle, by simple statistical reasoning. If you consider the various ways in which the components of a system—a collection of bricks, for example—can be arranged, you will find invariably that the possible random configurations are more numerous than the orderly ones. Thus disorder is favored, not because of some inherent sloppiness of the universe, but simply because disorder is more probable—that is, has more ways of being realized—than order.

But probability is a quantitative concept that can be evaluated. And so, therefore, must randomness be: some things are more random than others—bricks can be scattered, heaped, stacked, or assembled into a house. Thanks to the occurrence of such differences, random agitation can be converted into directed motion and made to work, as it is in a steam engine, provided more randomness is generated than is consumed in the process. Substitute the word entropy for randomness and you have the condition that limits the conversion of heat into kinetic energy or into any other kind of energy. Such a conversion is possible to the extent that the overall entropy increases or, at least, does not decrease. The italic letter S being the accepted symbol for entropy, this rule is expressed by the following equation, which is an alternative formulation of the second law:

$$\Delta S \geq 0 \qquad (4)$$

This equation tells us that entropy cannot decrease spontaneously. Order cannot arise out of chaos without help. The latter qualification is important. As with the ban against the increase of G, we must clearly define the limits of the system.

The relationship of entropy to heat is given by the following formula:

$$\Delta S = \frac{Q}{T} \qquad (5)$$

in which Q is a quantity of heat gained by a system, ΔS is the corresponding increase in entropy, and T the absolute temperature.

Without going into details, you can see from equation 5 that more entropy is gained by transferring a given quantity of heat at low than at high temperature. The trick, therefore, to making a heat engine is to remove heat from a hot source and transfer just enough of it to a cold collector to satisfy equation 4. The remainder can now be converted into work, provided you have arranged to make the transfer process contingent on such a conversion. A heat engine is a device of this sort; it is a machine in which the conversion of heat into work is harnessed to heat transfer. If Q is the heat given out by the hot source, Q' the heat gathered by the cold collector ($Q - Q'$ being the work obtained), and T and T' the absolute temperatures of the source and collector, respectively, you can easily find, by applying equations 4 and 5, that the fraction of the total heat produced that is converted into work is given by:

$$\text{Yield} = \frac{\text{work}}{\text{heat}} = \frac{Q - Q'}{Q} \leq \frac{T - T'}{T} \qquad (6)$$

Principle of heat engine. Of the Q calories delivered from the combustion chamber to the transducer at a temperature of T K, Q' are transferred to the cooler at T' K, and the difference $Q - Q'$ is converted into work. The maximum work obtainable is determined by the condition that

$$\Delta S = -\frac{Q}{T} + \frac{Q'}{T'} \geq 0$$

or

$$Q' \geq \frac{T'}{T} \times Q$$

For example, if T is 120°C (393 K) and T' is 20°C (293 K),

$$Q' \geq \frac{293}{393} \times Q = 0.75\, Q$$

At least 75 per cent of the heat delivered to the transducer must be returned to the cooler. As much as 25 per cent of this heat may be converted into work. If T is 300°C (573 K), almost 50 per cent of the heat can be converted into work.

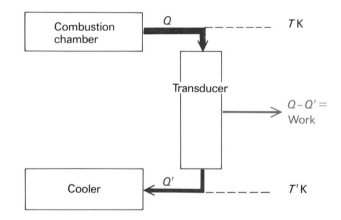

Thus, the yield, or efficiency, of a heat engine depends on the difference between the two temperatures. With T' limited to ambient temperature, you can gain only by increasing T. The internal combustion engine is superior to the steam engine in this respect. And Mr. Diesel achieved world fame by inventing an engine capable of running at a higher temperature than the ordinary internal combustion engine. On the other hand, without a temperature difference, there can be no conversion of heat into other forms of energy. A monothermic heat engine is impossible.

This statement disqualifies living systems as possible heat engines but not, however, as entropy engines. Randomness, it must be stressed, can be generated by many ways other than molecular agitation. Sugar is more random in a cup of coffee than in a lump; amino acids are more random in a mixture than

Two types of engine. In the internal energy engine, the system loses internal energy ($\Delta H < 0$) in the transformation from the initial state to the final state, and the energy lost is converted more or less completely into work. In the entropy engine, the internal energy content of the system does not change, but its entropy (degree of randomness) increases in the passage from the initial state to the final state. This transformation is harnessed to the conversion of external heat into work. The decrease ($\Delta S'$) in the entropy of the environment associated with this conversion is compensated by the increase (ΔS) in the entropy of the system. Thus the maximum work obtainable from external heat in this manner is $T\Delta S$ (total entropy change $\Delta S + \Delta S' = 0$).

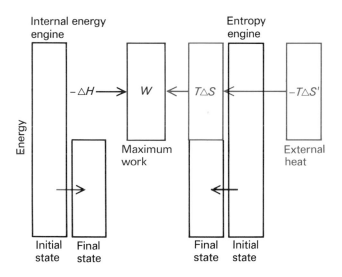

of system plus outside world is $\Delta S + \Delta S'$. According to equation 4, this total change in entropy cannot be negative. Therefore:

$$\Delta S + \Delta S' \geq 0 \qquad (7)$$

If ΔS is positive and $\Delta S'$ negative (as with the entropy engine), $T\Delta S$ is the maximum amount of outside heat that can be used for work. If ΔS is negative and $\Delta S'$ positive, $T\Delta S$ is the minimum amount of internal energy that the system must convert into heat and transfer to the outside.

If we combine the contributions of internal energy (enthalpy) and entropy to the capacity of a system to carry out work by undergoing a given transformation, we arrive at the definition of the free-energy change associated with the transformation (free energy of reaction):

$$\Delta G = \Delta H - T\Delta S \qquad (8)$$

As long as the entropy component is null or small, the free-energy change is about the same as the internal-energy change (internal-energy engine). But, if ΔS becomes appreciable, the relation between ΔG and ΔH changes to the point that a reaction can be exergonic (ΔG negative) in spite of being endothermic (ΔH positive), or endergonic in spite of being exothermic.

If the system undergoes the change without carrying out any work, ΔH is made entirely of heat given out to (ΔH negative) or received from (ΔH positive) the outside world. Then, $\Delta H = -T\Delta S'$, in which $\Delta S'$ is the entropy change of the outside world. Therefore, in the absence of work:

$$\Delta G = -T(\Delta S + \Delta S')$$

which, by virtue of equation 7, reduces to equation 1:

$$\Delta G \leq 0$$

As you can see, equations 1 and 4 indeed express the same principle. They are different formulations of the second law of thermodynamics. But beware of confusion. In equation 4, ΔS refers to the entropy change of a completely isolated system (or of the universe). In equation 8, on the other hand, ΔS represents the entropy change of a closed system capable of exchanging heat with the outside. To find out what happens to the entropy of the universe, we must include, as we have done, the change $\Delta S'$ in the entropy of the outside world.

A last remark. As already pointed out, the equality sign in every formula derived from the second law (see equations 1, 2, 4, 6, and 7) corresponds to an ideal limit situation. In the real

in a protein. An entropy engine is a device whereby a change that increases randomness—that is, has a positive ΔS—is coupled to the conversion of heat into work. If the system is isolated, it will cool, as does ether when it evaporates or a stretched rubber band that is allowed to contract. If the system is isothermic, as in our case, it will gain the corresponding amount of heat from the outside. Therefore, an isothermic entropy engine converts external heat into work, thanks to the increase of its internal entropy. These relationships work both ways. If the change undergone by the system has a negative ΔS, the entropy debt must be paid for by the conversion of internal energy into heat and its transfer to the outside world.

In connection with these heat exchanges, the entropy of the outside world suffers a change, $\Delta S'$, the amount of heat transferred being $T\Delta S'$ (equation 5). Thus, the total entropy change world, entropy always increases when something happens; the work we accomplish is always less than the energy we spend; some energy is always degraded to heat. There is an "entropy tax" levied on every energy transaction. This explains why time machines are impossible: the direction of entropy increase defines the arrow of time; there is no going back. Sometimes it is also said that the universe runs on its reserve of negative entropy, or *neguentropy*. When this runs out, nothing will happen any more, except statistical fluctuations. This gloomy prediction need hardly worry us, however. The sun alone has, we are told, another 5 billion years to go.

The Many Currencies of Energy and Work

Energy and work take many forms—most of them encountered in the living world—that are all interconvertible within the framework of the first law of thermodynamics, subject to the constraints imposed by the second law. Each form of work, or kind of energy, is defined in terms of its own internal parameters and evaluated in corresponding units.

Mechanical work is accomplished by a force that displaces its point of application a certain distance in the direction of action of the force. It is the product of the force times distance and is expressed in ergs (dynes × centimeters), joules (10^7 ergs), kilogram-meters (9.81 joules), foot-pounds (1.36 joules), or any other useful combination of the variables.

A related concept is *power*, which is the work performed, or energy consumed, per unit of time. The unit of power is the watt (one joule per second). The horsepower (550 foot-pounds per second, or 748 watts) and its French equivalent, the "cheval-vapeur" (75 kilogram-meters per second, or 736 watts, or 0.98 horsepower) illustrate the bewildering parochialism that still dominates the world's energy currency system.

Electric work depends on an electric charge moving across a potential difference and is given in terms of these two variables. It is expressed in joules (coulombs × volts) or in electron-volts (1.6×10^{-19} joules), which, multiplied by Avogadro's number ($N = 6.023 \times 10^{23}$), gives Faraday-volts (96,500 joules), useful in electrochemistry. Electric energy is often also expressed as power multiplied by time. The most commonly used such unit is the kilowatt-hour (3.6×10^6 joules).

Electromagnetic energy, including that of visible *light*, is defined by the number of quanta or photons and by their energy, which is a function of the frequency (ν) or wavelength (λ) of the light, according to the well-known equation of Planck:

$$\text{Energy of one photon} = h\nu = h\frac{c}{\lambda} \qquad (9)$$

in which h is Planck's constant (6.624×10^{-27} erg · sec), and c is the velocity of light (3×10^{10} centimeters per second). A number of photons equal to Avogadro's number is called an einstein. If wavelength is expressed in centimeters, the energy of one einstein is $12/\lambda$ joules.

The unit of heat, or *caloric energy*, is the calorie. It is the amount of heat required to raise the temperature of 1 gram of water from 15 to 16 degrees centigrade and is worth 4.186 joules. Its insular counterpart, the British thermal unit (BTU), raises the temperature of 1 pound of water at maximum density by 1 degree Fahrenheit. It equals 250 calories, or 1,046 joules.

Chemical energy in its various forms is traditionally expressed in calories, a heritage of thermochemistry and calorimetry. So is biological energy. Recently, however, the rulers of energy currency exchanges have decreed that henceforth the joule (symbol, J) is to be the official energy unit. But habits die hard, and few biochemists have as yet converted to the new system. This book is no exception. All energy values quoted in it are expressed in kilocalories (kcal). They can be readily converted into kilojoules (kJ) by multiplying by 4.186.

The Work of Molecular Transport

Translocation—by diffusion, for example—is the simplest molecular event that can occur, and it can serve as guide to more complex transformations. The associated free-energy change is evaluated by the change in the *chemical potential* of the substance undergoing translocation. Represented by μ, and expressed in kilocalories per gram-molecule (6.023×10^{23} molecules), the chemical potential is defined, in the legalistic jargon that must be used sometimes if rigor is to be respected, as the increase in free energy associated with the addition of an infinitesimally small amount of the substance (negligible with respect to the amount present) to an isothermic and isobaric system in which the substance is present at a defined partial pressure (if in gaseous phase) or concentration (if in solution).

Alternatively, and perhaps more intuitively, we may define the chemical potential as the increase in free energy associated with the addition of 1 gram-molecule of a given substance to a system—infinitely large, in fact—such that this addition does not appreciably alter the partial pressure, or concentration, of the substance in the system.

Both definitions are equivalent. They include the essential stipulation that the partial pressure or concentration of the substance must remain constant, because the chemical potential is a *function of the state of the system*. This relationship is given by an equation of absolutely fundamental importance, which dominates the whole field of chemical energetics:

$$\mu = \mu° + RT \ln a \qquad (10)$$

In this equation, $\mu°$ is a constant—characteristic for each substance—called the *standard chemical potential, R* is the gas constant (1.987×10^{-3} kcal per gram-molecule per degree), T is the absolute temperature in kelvins (K), ln is the natural logarithm ($2.3 \times$ the decadic logarithm), and a is the activity of the substance. At mammalian temperature ($37°C = 310K$), $RT \ln a = 1.417 \log a$, or, in first approximation, $1.4 \log a$.

Activity is another of those ideal limit concepts energetics is fond of. It is a measure of the effective abundance of the substance concerned, and corresponds exactly to its *partial pressure* (p) if it is a perfect gas, to its *concentration* (C) if it is an ideal solute. For real systems, it approaches partial pressure or concentration the more closely, the more rarefied the gas or dilute the solution. We will adopt these approximations for simplicity's sake and write:

$$\mu = \mu° + RT \ln p \qquad (11)$$

for a gas and:

$$\mu = \mu° + RT \ln C \qquad (12)$$

for a substance in solution.

The key message of these equations may be stated succinctly: *the more there is, the harder it is to add more.* Intuitively, we are not surprised. Still, there is something of a contradiction, considering that what we add (a certain number of molecules) is rigorously the same, whatever the partial pressure or the concentration. Therefore, the increase in the internal energy of the system associated with the addition should be—and is—the same under all circumstances. What is not the same, however, is the entropy change. The more molecules there are segregated in a given space, the greater the odds against an even greater degree of segregation, and the greater, therefore, in absolute value the (negative) entropy change associated with addition. This fact explains the activity-dependence of the chemical potential, expressed by equation 10. The chemical potential, it must be re-

membered, is a free-energy change, a function, therefore, of both internal energy and entropy, according to equation 8.

The logarithmic form of the second term of equations 10, 11, and 12 makes the increment in chemical potential additive for a multiplicative increase in abundance. A simple calculation shows that, at body temperature ($37°C$, 310 K), μ increases by about 0.42 kcal per gram-molecule for every doubling of partial pressure or concentration and by 1.4 kcal per gram-molecule for every tenfold increase in partial pressure or concentration. This increment is independent of the nature of the substance concerned—at least in the ideal world of theory, where molecules are taken to be either inert or too far apart to interact significantly with each other. In the real world things are different.

The nature of the substance does, however, enter into consideration with the standard chemical potential, $\mu°$, which is a characteristic of each substance. It is defined as the value of μ when the substance is in the *standard state*, which is itself defined by the state in which the activity, partial pressure, or concentration equals unity; that is, its logarithm is equal to zero. In this event, indeed, equations 10, 11, or 12 reduce to $\mu = \mu°$. Defining the standard state, therefore depends on the units we choose for expressing partial pressures or concentrations.

It is customary to express pressures in *atmospheres* and concentrations in *gram-molecules per liter* of solution (molarity). Therefore, $\mu°$ is the chemical potential of a gas at a partial pressure of 1 atmosphere, or of a solute at a concentration of 1 gram-molecule per liter (1 molar). We are perfectly free to choose other units if we so wish. This will simply mean that we give ourselves a different standard state and have to adjust the $\mu°$ values accordingly, so that the equation always gives us the same correct value of μ for a given partial pressure or concentration.

With this lengthy but necessary introduction we can now address our translocation problem. Imagine that a small amount of a substance passes from a compartment where its concentration is C_1 to one where its concentration is C_2. By definition of the chemical potential, the free energy of compartment 1 decreases by μ_1 kcal per gram-molecule, while that of compartment 2 increases by μ_2 kcal per gram-molecule. The overall free-energy change, expressed in the same units, is:

$$\Delta G = \mu_2 - \mu_1$$

or, by virtue of equation 12:

$$\Delta G = RT \, (\ln C_2 - \ln C_1) \qquad (13)$$
$$= RT \ln \frac{C_2}{C_1} \simeq 1.4 \log \frac{C_2}{C_1} \text{(at 37°C)}$$

It is the basic equation of molecular transport. Particularly simple because $\mu°$ cancels out in the operation, it tells in quantitive terms what we all know qualitatively from everyday experience—namely, that substances always move down concentration gradients. If C_1 is greater than C_2, ΔG is negative; transport can take place spontaneously. If C_1 is equal to C_2, $\Delta G = 0$; the system is at equilibrium. If C_1 is less than C_2, ΔG is positive; transport can occur only with the help of an external supply of energy equal at least to ΔG. It is the story of the sugar and the coffee and of countless other familiar phenomena. Both kinds of transport occur in living cells. They are called passive and active, depending on whether they can take place spontaneously or need to be driven. (Chapters 3, 13, and 14).

The Work of Ionic Transport

Equation 13 applies identically to electrically neutral and to electrically charged substances, or *ions* (within the limits of our ideal world; in the real world, ions of the same charge are less companionable than are neutral molecules because of the Coulomb repulsions between them). If an electric potential is interposed between the two compartments (membrane potential), a second term must be added to the free-energy equation to account for the electric work.

If z is the electric charge of the transported ion, then the total charge transferred with 1 gram-molecule of the ion is $z \times 96,500$ coulombs. If the membrane potential is represented by V (in volts), taken positively if compartment 2 is positive, then the electric work, in kilocalories per gram-molecule transported, is:

$$\text{Electric work} = \frac{96,500}{4,186} \cdot z \cdot V = 23 \, z \cdot V \qquad (14)$$

and (at 37°C):

$$\Delta G = 1.4 \log \frac{C_2}{C_1} + 23 \, z \cdot V \qquad (15)$$

The electric term may oppose transport (e.g., a positive ion moving against a positive potential) or it may help transport. Biology abounds in examples of both. The sodium-potassium pump, for instance (Chapter 13), forces Na^+ ions out of cells against a fifteen-fold difference in concentration ($+1.65$ kcal per equivalent) and against a membrane potential of the order of 0.07 volts ($+1.61$ kcal per equivalent), with a total cost of 3.26 kcal per equivalent. In contrast, the reverse transport of K^+ ions

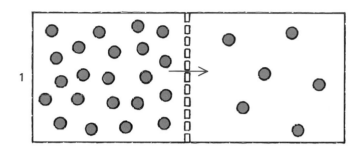

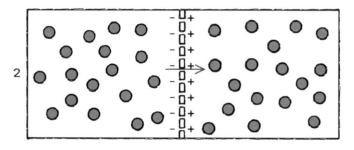

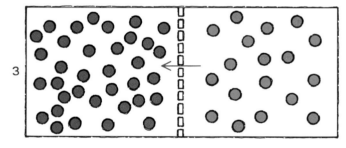

Some forms of chemical transport. Arrows show spontaneous direction ($\Delta G < 0$).

1. Simple molecular transport. A permeable membrane separates two regions of unequal chemical potential (partial pressure for gas or concentration for solute).

2. Ionic transport. In the example chosen, the chemical potential of the ion is the same in the two compartments. The sign of the membrane electric potential (V positive) determines the direction of movement of the ions, which are taken to be negatively charged (z negative).

3. Water transport through a semipermeable membrane impermeable to solutes. Total osmolalities on each side of the membrane determine the direction of water transport.

into the cells against a comparable gradient is helped by the membrane potential and costs essentially nothing. Or, more correctly stated, the active outward transport of sodium pays for itself and for the inward transport of potassium.

The Work of Water Transport

Theoretically, and also in practice, we can increase the concentration of a solution either by adding solute or by removing solvent. We tend to speak of *osmotic work* when water is transported rather than solute, as occurs, for instance, in the convoluted tubules of the kidneys.

The simplest form of osmotic work is accomplished when water is transported across a membrane that allows only water molecules, but none of the solute molecules, to pass through (semipermeable membrane). By the same reasoning as before, but applying equation 10 rather than equation 12, we find:

$$\Delta G = RT \ln \frac{a_2}{a_1} \tag{16}$$

in which a_1 and a_2 are the activities of water in the two compartments. It is customary in the thermodynamic analysis of solutions to express the activity of water by the *mole fraction*—that is, the number of molecules of water divided by the total number of molecules ($a = 1$ for pure water), which is related to the *osmolal* concentration of solutes, C', by:

$$a = \frac{55.56}{55.56 + C'} \tag{17}$$

In this formula, C' represents the sum of the molal concentrations of all the solutes present. Molal concentrations are expressed in gram-molecules per 1000 grams (55.56 gram-molecules) of water, and not per liter of solution (molarity), as we do elsewhere. For dilute solutions, molality and molarity are not very different. The prefix *os* (from osmotic) indicates that all concentrations have been added up.

Introducing equation 17 into equation 16, we find:

$$\Delta G = RT \ln \frac{55.56 + C'_1}{55.56 + C'_2} \tag{18}$$

which represents the free-energy change, in kilocalories per gram-molecule of water transported, associated with the transport of water from a compartment of osmolality C'_1 to one of osmolality C'_2.

It is easy to see that water transport is not a costly business in biological systems. For example, returning 1 liter of water from concentrated urine (1.3 osmolal) to blood (0.3 osmolal) requires no more than 0.6 kcal, which, therefore, also represents the amount of free energy that would be released by 1 liter of water flowing freely from urine to blood through a semipermeable membrane. Surprisingly, the pressure that would have to be applied on the blood side to oppose water transfer (*osmotic pressure*) is very large, of the order of 25 atmospheres. The paradox is only apparent, of course: 25 liter-atmospheres are, in fact, equal to 0.6 kcal.

In conclusion, large differences in osmotic pressure, such as those that keep the vacuoles of plant cells (tonoplasts) under tension and maintain turgidity, are not expensive to generate. The fundamental reason for this is that, because of the great abundance of water with respect to solutes in most biological systems, its thermodynamic activity suffers only small variations and remains close to that of pure water. In most bioenergetic calculations, the activity of water is, in fact, assumed to be constant and equal to unity (see p. 422).

Chemical Work

Formally, a chemical transformation can be treated in terms of translocation. Take the simple reaction:

$$A \rightleftharpoons B$$

From left to right, it is equivalent to adding molecules of B to the system and removing (adding negatively) the same number of molecules of A from the system. Therefore, by virtue of the definition of the chemical potential, the free-energy change associated with the transformation (*free energy of reaction*), expressed in kilocalories per gram-molecule of A consumed or of B formed, may be formulated as follows:

$$\Delta G = \mu_B - \mu_A$$

or, according to equation 12 (assuming the reaction takes place in solution):

$$\Delta G = \mu°_B - \mu°_A + RT (\ln [B] - \ln [A])$$

in which brackets are used to express concentrations. This relationship can be rewritten:

$$\Delta G = \Delta G° + RT \ln \frac{[B]}{[A]} \tag{19}$$

Chemical reaction. Formally, a chemical reaction may be treated as chemical transport from and into outside reservoirs. In the example shown, B enters the system and A leaves it in identical amounts. Therefore, for the system itself (central compartment):

$$\Delta G = \mu_B - \mu_A$$

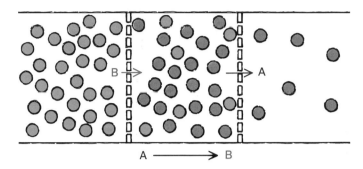

$$A \longrightarrow B$$

in which $\Delta G° = \mu°_B - \mu°_A = $ *the standard free energy of the reaction.*

We could consider the reaction from right to left in the same manner and would find exactly the same value of ΔG, but with the opposite sign. Therefore, there is necessarily a direction for which ΔG is negative; it is the exergonic, spontaneous direction. The reverse direction is endergonic and cannot be taken without an external supply of free energy. Note, however, that these directions are not immutable. Even if $\Delta G°$ is positive, ΔG can still be negative, and the spontaneous direction can be from left to right. All that is needed is for the logarithmic term in equation 19 to be negative ($[B] < [A]$) and greater in absolute value than $\Delta G°$. In contrast (assuming again that $\Delta G°$ is positive), the reaction will proceed spontaneously from right to left if the logarithmic term is positive ($[B] > [A]$) or if it is negative and smaller in absolute value than $\Delta G°$. If the two terms cancel each other out, $\Delta G = 0$, the system is at equilibrium.

We encounter the same entropy factor as we did in simple transport. The internal energy bill of making B from A is the same, whatever the state of the system. Not so the entropy bill. The more of B that is already present, the harder it is to make more, and vice versa.

By definition, $\Delta G°$ ($=\mu°_B - \mu°_A$) is the free energy of the reaction when both A and B are in the standard state. According to equation 19, it is also the free energy of the reaction in all situations in which the logarithmic term is equal to zero—that is, whenever $[A] = [B]$.

Note, finally, that at equilibrium ($\Delta G = 0$), we have:

$$\Delta G° = -RT \ln\left(\frac{[B]}{[A]}\right)_{eq}$$

But we know from chemical kinetics (law of mass action)

that the *equilibrium constant*, K, of the reaction is none other than:

$$K = \left(\frac{[B]}{[A]}\right)_{eq}$$

Therefore:

$$\Delta G° = -RT \ln K \qquad (20)$$

This is an important relationship, which underlies a major method of determining free energies of reaction through the equilibrium constant. In our example, all we need to know are the concentrations of A and B at equilibrium.

The preceding notions can be generalized to every possible chemical transformation. Consider the following, entirely general, formulation of a chemical reaction:

$$m_A\ A + m_B\ B + m_C\ C + \cdots \rightleftharpoons$$
$$m_M\ M + m_N\ N + m_Q\ Q + \cdots$$

Extending our earlier reasoning, we readily find:

$$\Delta G = m_M\ \mu_M + m_N\ \mu_N + m_Q\ \mu_Q +$$
$$\cdots - m_A\ \mu_A - m_B\ \mu_B - m_C\ \mu_C - \cdots$$

or, using the capital Greek sigma, Σ, for sum and representing the products (M, N, Q, . . .) of the reaction by P and the reactants (A, B, C, . . .) by R:

$$\Delta G = \Sigma m_P\ \mu_P - \Sigma m_R\ \mu_R \qquad (21)$$

Replacing the chemical potentials μ by their values (equation 13) and rearranging the terms of the equation according to the rules of logarithmic calculation, we obtain:

$$\Delta G = \Delta G° + RT \ln \frac{\Pi[P]^{m_P}}{\Pi[R]^{m_R}} \qquad (22)$$

in which:

$$\Delta G° = \Sigma m_P\ \mu°_P - \Sigma m_R\ \mu°_R \qquad (23)$$

and the capital Greek pi, Π, stands for product.

Equation 22 is the generalized form of equation 19. It reduces identically to equation 20 at equilibrium. In it, ΔG and $\Delta G°$ are expressed in kilocalories per m_A gram-molecules of A, or m_B of B, m_C of C, etcetera, consumed, or per m_M gram-molecules of M, or m_N of N, m_Q of Q, etcetera, formed—that is, in kilocalories per *stoichiometric equivalent* (Greek *stoikheion*, element) of reactant consumed or of product formed.

Living cells are chemical machines; they carry out thousands of reactions, many of which play key roles in providing free energy, or using it, for the performance of work through coupled systems or transducers. If we wish to understand these mechanisms, which is the object of bioenergetics, we must know the ΔGs of the reactions. For this purpose, biochemists have determined a large number of $\Delta G°$ values from equilibrium studies (equation 20) or by other methods. Lists of these values can be found in most textbooks, usually corrected for hydrogen-ion concentration, which, in living cells, is of the order of 10^{-7} equivalents per liter (pH 7.0), or 7 orders of magnitude away from the standard state, corresponding to a decrease in chemical potential by $1.4 \times 7 = 9.8$ kcal per equivalent.

Even so corrected, these $\Delta G°'$ values are rarely directly applicable to physiological situations because the concentrations of metabolites and cofactors in living cells are all much lower than standard. They range from about 0.01 gram-molecule per liter, for the most abundant substances, to concentrations that may be as low as 10^{-6} gram-molecule per liter or lower. Furthermore, these concentrations vary according to circumstances; in addition, they may require fairly extensive corrections before they can be substituted for the activities of the substances because the situation in living cells is very different from that of the very dilute solution that must be assumed if we are to use equation 12 instead of the rigorously correct equation 10.

For these reasons it is extremely difficult to know any physiological ΔG value accurately. But in many cases an estimate of the normal range is available. The ΔG values quoted in this book, referred to as "physiological," are based on such estimates, as is explained at the end of Chapter 7 and elsewhere.

The Work of Electron Transfer

The most striking revelation of bioenergetics has been the universal importance of electron transfer as almost exclusive purveyor of energy throughout the living world. Such reactions also occur in the nonliving world. Their analysis, following the development of *electric cells*, provided classical thermodynamics with a new important chapter, which turned out to be particularly useful for the understanding of biological mechanisms.

Consider what happens when metallic zinc is exposed to sulfuric acid:

$$Zn + H_2SO_4 \longrightarrow ZnSO_4 + H_2\nearrow$$

The zinc dissolves as zinc sulfate and hydrogen gas is evolved.

Actually, the sulfate ion does not participate in the reaction, which is really a transfer of electrons from zinc atoms to hydrogen ions:

$$Zn + 2\,H^+ \longrightarrow Zn^{2+} + H_2$$

This becomes particularly clear when the preceding reaction is broken down formally into two half-reactions:

$$Zn \longrightarrow Zn^{2+} + 2\,e^-$$
$$2\,H^+ + 2\,e^- \longrightarrow H_2$$

By definition—a historical consequence of the powerful electron-stripping ability of oxygen—removal of electrons, as undergone by zinc, is called an *oxidation*; acquisition of electrons, as by the hydrogen ions, is a *reduction*. Electron transfer is an *oxidation-reduction*.

The reaction between zinc and protons is exothermic. Under standard conditions, it produces some 35 kcal per gram-atom of zinc oxidized (65 grams) or per pair of proton-equivalents reduced (2 grams). Most of this energy is free energy ($\Delta H \simeq \Delta G$); it need not appear as heat and could be used for work. This is achieved by the physical separation of the two half-reactions in two distinct half-cells in such a manner that the electrons delivered by the zinc can reach the hydrogen ions only through an outside conductor. Any electric transducer can then be interposed in the pathway of the electrons, thereby coupling electron transfer to the performance of work. All electric cells function on this principle.

The maximum amount of work, W, that can be accomplished is readily computed from the difference in electric potential, E (volts), existing between the zinc terminal and the hydrogen terminal. Expressed in kilocalories (4,186 J) per pair of electron-equivalents ($2\,\mathscr{F} = 2$ Faradays $= 2 \times 96,500$ coulombs):

$$W = -\frac{2\,\mathscr{F}}{4,186}\,E = -46\,E \tag{24}$$

By definition, $W = -\Delta G$. In consideration of equation 22, we find:

$$-W = \Delta G = \Delta G° + RT \ln \frac{[Zn^{2+}]\,p_{H_2}}{[Zn]\,[H^+]^2} \tag{25}$$

In the particular case where the components of the hydrogen half-cell are in the *standard state* ($p_{H_2} = 1$ atmosphere; $[H^+] = 1$ equivalent per liter), we obtain:

$$-W = \Delta G = \Delta G° + RT \ln \frac{[Zn^{2+}]}{[Zn]} \tag{26}$$

Schematic representation of electron transfer in an electrochemical cell. The transfer of an electron from A⁻ to B can take place only by way of the electrodes and of the outside con-

ductor. The free energy of the transfer can be harnessed to do work. It can be estimated from the electric potential difference of the cell.

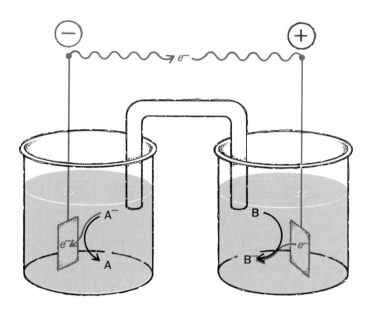

It is the Nernst equation.

One of the main advantages of oxidation-reduction potentials is that they allow a very easy determination of the ΔG for any electron-transfer reaction between redox couples whose standard oxidation-reduction potentials are known. Consider the following general reaction:

$$A_{red} + B_{ox} \rightleftharpoons A_{ox} + B_{red}$$

which decomposes to:

$$A_{red} \rightleftharpoons A_{ox} + n\,e^-$$
$$B_{ox} + n\,e^- \rightleftharpoons B_{red}$$

From left to right, in consideration of equations 24 and 25:

$$\Delta G = \frac{n\,\mathscr{F}}{4,186}\,E_{(A-B)} \tag{30}$$

in which $E_{(A-B)}$ is the potential difference that would exist between the A and B half-cells should we build an electric cell with these two couples. Actually, we need not do that if we know the oxidation-reduction potentials, E_A and E_B, of A and B. Then, for obvious reasons (both E_A and E_B have been measured against the same standard hydrogen half-cell):

$$E_{(A-B)} = E_A - E_B$$

or, by virtue of equation 29:

$$E_{(A-B)} = E°_A - E°_B + \frac{R'T}{n\,\mathscr{F}} \ln \frac{[A_{ox}]\,[B_{red}]}{[A_{red}]\,[B_{ox}]}$$

which, combined with equation 30, gives:

$$\Delta G = \frac{n\,\mathscr{F}}{4,186}(E°_A - E°_B) + RT \ln \frac{[A_{ox}]\,[B_{red}]}{[A_{red}]\,[B_{ox}]}$$

It follows, in view of equation 22, that:

$$\Delta G° = \frac{n\,\mathscr{F}}{4,186}(E°_A - E°_B) = n \cdot 23 \cdot (E°_A - E°_B) \tag{31}$$

Standard oxidation-reduction potentials are known for most important biological redox couples. They are listed in textbooks, generally as $E°'$ (i.e., corrected for the hydrogen-ion concentration, as are $\Delta G°'$ values, p. 420).

In bioenergetics, it is often useful to know the total amount of free energy that can be made available by the oxidation (defined as the removal of one pair of electrons) of a given substance, with atmospheric oxygen as electron acceptor. This is estimated from the oxidation-reduction potential of oxygen,

which, combined with equation 24, gives:

$$E = \frac{4,186\,\Delta G°}{2\,\mathscr{F}} + \frac{4,186\,RT}{2\,\mathscr{F}} \ln \frac{[Zn^{2+}]}{[Zn]} \tag{27}$$

or:

$$E = E° + \frac{R'T}{2\,\mathscr{F}} \ln \frac{[Zn^{2+}]}{[Zn]} \tag{28}$$

in which R' is the gas constant expressed in joules per degree per gram-molecule. (Earlier, the gas constant R was expressed in kilocalories per gram-molecule per degree; $R' = 4,186\,R = 8.317$.) Note that [Zn] does not have to figure explicitly in this equation because the thermodynamic activity of a solid metal is equal to unity under all conditions.

By definition (i.e., when measured against a standard hydrogen half-cell), E, as given by equation 28, is the *oxidation-reduction potential* of the Zn/Zn^{2+} couple, and $E°$ is its standard oxidation-reduction potential. In general, for any *redox couple*:

$$Red \rightleftharpoons Ox + n\,e^-$$

Equation 28 becomes:

$$E = E° + \frac{R'T}{n\,\mathscr{F}} \ln \frac{[Ox]}{[Red]} \tag{29}$$

which, under "physiological" conditions, is of the order of +0.805 volts. Therefore:

$$\Delta G_{ox} = \frac{2 \mathscr{F}}{4,186} (E - 0.805) = 46 E - 37 \qquad (32)$$

in which E is the "physiological" oxidation-reduction potential of the redox couple under consideration. In Chapter 7, ΔG_{ox} is referred to as the *electron potential*, not to be confused with oxidation-reduction potential (it is measured in kilocalories per pair of electron-equivalents, not in volts).

The Work of Generating Electricity

Electricity is always created in biological systems through an unequal distribution of electric charges on the two sides of a membrane. The maintenance of this unequal charge distribution is ensured by the lipid bilayer, and inserted proteins make up the necessary channels, gates, and pumps (Chapters 3 and 13). One can imagine charge separation to be caused either by electron transport or by ion transport. In practice, it seems that only ion transport is directly involved, although the two processes may be intimately linked, as in the chemiosmotic transducers of mitochondria (Chapter 9) and chloroplasts (Chapter 10).

By definition, an *electrogenic* ion pump is an energy-driven transport system that translocates ions across a membrane impermeable to these ions (otherwise the pump is short-circuited) without built-in charge compensation (such as the obligatory accompaniment by a counter-ion of opposite sign, or exchange for an ion of the same sign). An electrogenic pump may, however, drive the transport of a compensating ion by means of the membrane potential it generates. For this to happen, the membrane must contain a gate or a carrier (ionophore) for the ion transported in this way (Chapters 9 and 13).

The main electrogenic pumps transport sodium ions or protons; but there may be others. The potentials they generate vary from a range of 50 to 70 mV (plasma membrane) up to as much as 200 mV or more (mitochondria, chloroplasts). *Currents* are produced as a result of a local disturbance that renders the membrane temporarily permeable to the electrogenic ion. Depolarization is transmitted laterally along the membrane—of an axon, for instance—by some sort of domino effect. In the electric organs of certain fishes, such as *Torpedo*, or the electric eel, *Gymnotus*, electricity-generating cells (electroplaxes), are organized so as to link their potentials in series in the form of batteries capable of producing discharges of several hundred volts.

The energetic cost of these various activities is paid in the currency of ion transport (see p. 416).

The Work of Biosynthesis

As explained in Chapter 8, most of the innumerable endergonic assembly reactions involved in the construction of the complex constituents of living cells are driven by exergonic processes to which they are coupled by *group-transfer* reactions. The energy bookkeeping of such processes is done by means of the *group potentials*—that is, the free energies of hydrolysis, ΔG_{hy}, of the substances involved.

Consider the following B-transfer reaction:

$$A{-}B + CH \rightleftharpoons AH + B{-}C$$

We decompose it formally into the hydrolysis of A—B and the dehydrating condensation (reversal of hydrolysis) of B—C:

$$A{-}B + H_2O \rightleftharpoons AH + BOH$$
$$B{-}OH + CH \rightleftharpoons B{-}C + H_2O$$

and write:

$$\Delta G_{(transfer)} = \Delta G_{hy\ (A{-}B)} - \Delta G_{hy\ (B{-}C)}$$

This elementary accounting system can be extended to sequential group transfers and coupled processes of any sort of complexity. It also serves to weigh group transfer against electron transfer and to quantitate oxidative phosphorylation. Most biologically important group-potential values have been determined. They are usually listed as $\Delta G^{\circ\prime}_{hy}$, or standard free energies of hydrolysis corrected to physiological pH.

Even so corrected, such values can be misleading. The value of $\Delta G^{\circ\prime}_{hy}$ for the hydrolysis of ATP to ADP and inorganic phosphate is of the order of -7.0 kcal per gram-molecule. The real value, in the mitochondria of living cells, may reach twice this amount (Chapter 9). This is because the [ATP] to [ADP] ratio may exceed 100, and the concentration of inorganic phosphate, $[P_i]$, may fall as low as 0.001 gram-molecule per liter. Introduced into equation 22, these values give:

$$\Delta G = \Delta G^{\circ\prime} + RT \ln \frac{[ADP]\,[P_i]}{[ATP]}$$

$$= -7.0 + 1.4 \log \frac{0.001}{100} = -14.0$$

Note that the concentration (activity) of water is not taken into account explicitly in the computation of group potentials, even though water is the key reactant in all hydrolysis reactions. As discussed earlier (p. 417), the activity of water varies very little in biological systems. In view of this fact, it is assumed to be invariable, at a value close to the activity of pure water, and is included in the $\Delta G^{\circ\prime}$ term. This is legitimate as long as we are dealing with aqueous systems. Things are different in nonaqueous media. For hydrolysis reactions taking place in an oil droplet or in the lipid bilayer of a membrane, water may become a critical reactant in determining the direction of the reaction.

Photochemical Work

The conversion of light energy into work always requires as first step the *absorption* of a quantum of light, or photon, by a molecule, which is said to become *excited* in the process. Molecules can store energy in four different forms: linear motion (translation), rotation, vibration, and electronic excitation. The last three can vary only by discontinuous jumps (quanta) and are involved in the absorption of light. The rule is that a photon can be absorbed only if its energy content, $h\nu$, is such as to allow the absorbing molecule to jump to an authorized energy level. These energy levels are readily identified from the *absorption spectrum*. Each absorption band corresponds to an energy jump equal to $28,600/\lambda$ kcal per gram-molecule (equation 9, but with λ expressed in nanometers).

Compared with the ΔG range of biochemical reactions, the amounts of energy that molecules equipped with the appropriate light-absorbing chemical groupings (*chromophores*) can collect from sunlight are considerable. For visible light ($\lambda = 800{-}400$ nm), they vary between 36 and 72 kcal per gram-molecule. They exceed 100 kcal per gram-molecule in the ultraviolet region, where proteins ($\lambda = 280$ nm) or nucleic acids ($\lambda = 260$ nm) have absorption bands. There is, indeed, great power in sunlight, provided you can catch it and channel it through an effective transducer.

If the transducer is photochemical, it has to act fast because molecules do not remain excited very long. Usually, they lose their excitation energy at the first collision with another molecule and convert it into translation energy. All we are left with is useless heat. Sometimes—this happens with special chromophores called fluorophores—part of the absorbed light is re-emitted in the form of light of higher wavelength. This phenomenon, called *fluorescence*, is used on a large scale to convert "cold" ultraviolet light into visible light. It is also widely used in the study of cells (Chapter 12). In certain favorable arrangements, as in the chloroplast thylakoids (Chapter 10), the excited state may be transferred to a neighboring molecule with little loss of utilizable free energy.

In a relatively small number of cases, absorbed light may serve to drive a photochemical reaction. Most often this reaction is exergonic, and the role of light is simply to help molecules jump over the energy barrier (activation energy) that prevents the reaction from taking place (*photocatalysis*). Some photocatalytic reactions are chain reactions and have an explosive character. Sometimes the photochemical reaction is actually endergonic, and light provides the necessary ΔG. The fundamental reaction of photosynthesis is of this kind (Chapter 10). With photons of sufficiently high energy, molecules may break apart (excessive vibration) or lose electrons (excessive electronic excitation). The *free radicals* generated in this manner are highly reactive and may cause many transformations, including mutations.

Some photochemical transducers operate in the reverse direction. The molecules are excited chemically with the help of an exergonic reaction and fall back to ground level by emitting photons. Molecules of this kind are involved in biological *luminescence*.

The Work of Information

The Scottish physicist James Clerk Maxwell, the founder of the electromagnetic theory of light, is also remembered for throwing an embarrassing pebble in the entropy pond. "Imagine," he said, "two adjacent compartments filled with the same gas at the same temperature, communicating with each other by a small opening, where stands a 'demon' who allows only fast-moving molecules to pass in one direction, and slow-moving ones in the other. Soon, one compartment will be warmer, the other will be colder. Entropy will have decreased, order arisen out of disorder, with no other help than *information* at the molecular level."

It took almost a century before the paradox was solved: the demon, in order to obtain the information he needs, consumes neguentropy. There is a relationship between information and entropy, which is readily grasped when both concepts are expressed in terms of probability. Entropy, as we have seen, is a measure of randomness, which itself depends on the probability of the particular configuration considered. According to statistical mechanics, the relationship is a simple logarithmic one and is given by:

$$\Delta S = k \ln \mathscr{P} \qquad (33)$$

in which ΔS represent the entropy change associated with the creation of a configuration of probability $\mathscr{P}$, and k is the Boltzmann constant ($R/N = 3.3 \times 10^{-27}$ kcal per degree).

Information, on the other hand, is measured in binary units, or *bits*, the number of $+/-$ choices needed for its expression. If we represent information content by I, we may write:

$$\mathscr{P} = 2^{-I} \qquad (34)$$

or:

$$I = -\log_2 \mathscr{P} = -1.45 \ln \mathscr{P} \qquad (35)$$

Hence, according to equation 33:

$$\Delta S = -0.69 \, kI \qquad (36)$$

As an example, take the information content of the human haploid genome (Chapter 16). With 3×10^9 base pairs, and 2 bits per base pair, $I = 6 \times 10^9$ bits. Hence, $\Delta S = -13.7 \times 10^{-18}$ kcal per degree, and the minimum work ($-T\Delta S$) in instructing the synthesis of one molecule at body temperature is 4.24×10^{-15} kcal, or 2.55×10^9 kcal per gram-molecule. This can be compared with the minimum chemical work involved in assembling 3×10^9 pairs of nucleotides into a double-stranded DNA ($\Delta G_{hy} \simeq -12$ kcal per gram-molecule for the phosphodiester bonds of DNA), or $3 \times 10^9 \times 2 \times 12 = 72 \times 10^9$ kcal per gram-molecule. The cost of information is small with respect to that of assembly, but not negligible.

The Work of Not Changing

If a Ping-Pong ball is at rest on a table, no work is required to keep it there. But, if it is floating in air, we must perform work in a continuous fashion—by means of a water jet, for example—to prevent it from falling. Living organisms are, like the floating ball, away from equilibrium and require a continuous expenditure of energy to be maintained in this unstable *steady state*. This kind of work obviously depends on the intrinsic instability of the systems considered. The sodium-potassium pump must work harder to maintain a sodium-ion gradient across a leaky membrane than across a tight membrane. A cell with a high degree of autophagy must carry out more biosynthetic repair than a cell that engages in little self-destruction.

When pure maintenance work is performed, nothing actually changes in the system concerned. Therefore, the work consists simply in the conversion of a given amount of free energy into heat. It is given per unit of time (t) and is expressed as the *flux* of free energy supplied to the system, $\Delta G/\Delta t$, or as the flux of entropy, $\Delta S/\Delta t$, generated in the system, which is equal to the free-energy flux divided by the absolute temperature.

A new branch of thermodynamics, called nonequilibrium thermodynamics, or thermodynamics of irreversible processes, has been developed for the formal analysis of systems traversed by such fluxes. In practice, the fluxes themselves can be measured by determining the rate of heat production under conditions in which the systems may be assumed to perform no work other than maintenance work. Physicians try to approximate such conditions when they measure the *basal metabolic rate*. For an adult human being, it amounts to about 2,000 kcal per day, which is equivalent to the consumption of a 100-watt bulb. That is what each of us has to use up simply to keep alive.

Sources of Illustrations

page 329
Electron micrograph courtesy of Dr. David
S. Hogness.

page 333
After Dr. R.W. Hart.

page 334
Photographs courtesy of Dr. Karl F. Koop-
man, American Museum of Natural History.

page 338
Electron micrograph courtesy of Dr. David
Dressler and Dr. Huntington Potter, Har-
vard University.

page 352
Scanning electron micrographs courtesy of
Dr. Marcel Bessis.

page 354
Phylogenetic tree courtesy of Dr. Emanuel
Margoliash.

page 362 (right)
Courtesy of Dr. Ralph L. Brinster.

page 370 (right)
Electron micrograph courtesy of Dr. Ulrich
K. Laemmli.

page 371
Karyotype courtesy of Dr. Herman Van
den Berghe.

page 372
Electron micrograph courtesy of Dr.
Etienne de Harven.

Index

This index includes entries from both volumes. Page numbers in boldface type refer to illustrations; those in italic type refer to definitions, etymologies, and structural diagrams.

Ecdysone, **317**, 318

Echinosphaerium nucleofilum, **210–211**

EcoRI, 334, **335**

Effectomers, 234

Efficiency, 410, 412

EGF (epidermal growth factor), 365, 367

Egg cells, 61, 92, 293, **295**, **302–304**, 336, 378, 382

  cytoplasm of, 315

  fertilization of, 287, 313–314, 377, 382

Ehrlich, Paul, 47, **48**, 250

Einstein, Albert, 18, 357, 409

Elastin, *40*

Electric cell, 419–420

Electric fishes, 421

Electricity

  generation of, 421

  work produced, 112, *414*, 416, 420

  See also Bioelectricity

Electrochemical potential, 227–230

Electron microscopy, 82, 193, 222, 292, 293, 308, 324

  development of, 11

Electron potential, 117–118, 122, 154, 159, 177, 421

Electron transfer, 111–118, 130, 239, 246, 397–402

  acceptors, 112, 114, 115–116, 177, 182, 188

  in autotrophs, 167–179

  energetics of, 419–421

  in glycolysis, 112–115

  in mitochondria, 153–165

  overflow mechanism, 179

  reversed, 118, 162–163, 179

Electrons

  destabilization of, 177

  high-energy, 240

  photoactivation of, 171–173, 177

Electrophilic attack, 122

Electrophoresis, **13**, **14**

Electroplaxes, 421

Elongation factors, 268–269

Embryos, 314

Enantiomorphs, *386*

Endergonic processes, *108*, 122, *409–410*, 413, 418, 421, 422

Endocytosis, 20, *53–59*, **65**, 77, 81, 210, 223, 230, 274

  receptor-mediated, 55–59, 95, 218–222, 233–234

  role of, 61

  See also Phagocytosis; Pinocytosis

Endonucleases, 332, **337**

Endoplasmic reticulum, **69**, *82–87*, 92–99, 152, 187, 223, 224, 274

  biosynthesis in, 237–238, 265, *405*

  and nuclear envelope, 280, 284, 374

  pink patches of, 239

  rough-surfaced, **13**, **14**, 20, **83**, 86, **88**, 97, **105**, **225**, 271–273, 275, 383

  smooth-surfaced, **13**, 20, **83**, 86, **88**, 383

  tubular connections, 104, 105

β-Endorphin, 275

Endosomes, 20, 55, 58, 61, **65**, 90, 95, 97, 234

  proton trapping, 76, 163

  viruses in, 103

Endosymbiont hypothesis, 149–150, 152, 166, 173–174, 175, 179, 187, 190, 209, 259, 271

Endothelial cells, *35*, 61

Endothermic transformations, 411, 413

Endotoxins, 75

Energetics, 20, 49–50, 106–108, 242, 408–423

  of biosynthesis, 120–147, 421–422

  of electricity generation, 421

  of electron transfer, 419–421

  of glucose production, 232

  of glycolysis, 106–118, 153, 155

  of hydrogen formation, 188

  of ionic transport, 416–417

  of mitochondrial oxidations, 153–165

  of molecular transport, 49–50, 414–416

  of nuclear biosynthesis, 283, 298, 300

  of photosynthesis, 177–179

  of sodium-potassium pump, 228–229

  of water transport, 417

Energy

  activation, 422

  biological, 414

  caloric, 414

  chemical, 414

  conservation of, 409

  degradation of, 409, 411, 414

  electric, 414

  electromagnetic, 414

  excitation, 422

  flux of, 423

  internal, 411, 413, 415

  of light, 177, *414*, 422

  of reaction, 413

  storage of, 154

  translation, 422

  units of measurement of, 414

Engines

  diesel, 412

  entropy, 412–413

  heat, 412

  internal combustion, 412

  steam, 408, 411, 412

Enkephalin, 275

Enthalpy, *411*, 413

Entropy, 244, *411–414*, 415, 418

  and heat, 412

  and information, 422–423

Enzymes, *28*, 36

  activation of, 223, 232

  binding sites for, 345

  biosynthesis of, 79, 81

  covalent modification of, 248

  digestion-resistent, 72

  and genetic code translation, 263–265

  kinetics of, 245–249

  lock-key analogy, 250

  lysosomal, **65**, 67, 81, 90 92, 237, 239

  lytic, 79–80, 81, 275

  mammalian, 351

  membrane-dissolving, 102

  mitochondrial, 153

  multienzyme complexes, 141

  oxidizing, 401

  peroxisomal, 182, **186–187**

  photoreactivating, 332

  specificity of, 108–109, 120, 243

  substrate-regulated, 248, 249

  See also specific enzymes and types of enzymes

Eobacterium isolatum, 106–107

Epidemiology, 363

Epidermal growth factor (EGF), 233

Epinephrine, 80, 230, 231–232, 249

Epithelial cells, 233

  differentiation of, 193–195

Equilibrium

  chemical, 418–419

  thermodynamic, 409, 410, 423

Erythrocytes, **2**, **34**, *35*, 352

  phagocytosis of, **54**